Aeroform

Aeroform: Designing for Wind and Air Movement provides a comprehensive introduction to applying aerodynamic principles to architectural design. It presents a challenge to architects and architectural engineers to give shape to the wind and express its influence on architectural form.

The wind pushes and pulls on our buildings, infiltrates and exfiltrates through cracks and openings, and lifts roofs during storm events. It can also offer opportunities for resource conservation through natural ventilation or a biophilic connection between indoors and out. This book provides basic concepts in fluid mechanics such as materials, forces, equilibrium, pressure, and hydrostatics; introduces the reader to the concept of airflow; and provides strategies for designing for wind resistance, especially in preventing uplift. Natural ventilation and forced airflow are explored using examples such as Thomas Herzog's Hall 26 in Hanover, RWE Ag building in Essen Germany, and the Kimbell Art Museum in Texas. Finally, issues of wind and airflow measurement are addressed.

A reference for students and practitioners of architecture and architectural engineering, this book is richly illustrated and presents complex concepts of aerodynamic engineering in easy-to-understand language. It prepares the architect or architectural engineer to design buildings that are visually expressive of a dialogue between wind and built form.

James Jones is Professor of Architecture in the School of Architecture + Design (SA+D) at Virginia Tech. He received his BS, MArch, and PhD from the College of Architecture and Urban Planning at the University of Michigan. He provides instruction related to environmental building systems, resource conservation, and health and wellness design to undergraduate and graduate students. Dr. Jones oversees the design research stream within the PhD Program for the SA+D. He also directs the Center for High Performance Environments, which has been recognized as a Center of Excellence by the US Department of Energy and the Environmental Protection Agency. He has been a Continuing Education Provider for the American Institute of Architects and a member of the American Society of Heating, Refrigerating and Air-Conditioning Engineers. He was Principle Investigator for the design development and patent of the V2T roof vent system.

Demetri Telionis is the Francis J. Maher emeritus professor at Virginia Tech. He received his MS and PhD from Cornell University and then served on the faculty of Virginia Tech. Telionis directed and participated in research funded by National Aeronautics and Space Administration, National Science Foundation, Oakridge National Lab, Air Force Office of Scientific Research, Department of Energy, Naval Air Systems Command, Boeing Corp. and Department of Transportation. He has written one book, edited three more, directed the PhD research of fifteen students, and authored or co-authored over 160 papers. Dr. Telionis is a fellow of the American Society of Mechanical Engineers, has served for three years as an associate editor of the *ASME Journal of Fluids Engineering*, and then for another ten years as the editor in chief of this journal. Telionis is one of the founders of Aeroprobe Corp, a company dedicated to the design and marketing of engineering instrumentation and has served three years as its CEO.

Aeroform
Designing for Wind and
Air Movement

James Jones and Demetri Telionis

Routledge
Taylor & Francis Group

NEW YORK AND LONDON

Cover image: Hall 26, Dieter Leistner

First published 2023
by Routledge
605 Third Avenue, New York, NY 10158

and by Routledge
4 Park Square, Milton Park, Abingdon, Oxon, OX14 4RN

Routledge is an imprint of the Taylor & Francis Group, an informa business

Library of Congress Cataloging-in-Publication Data
Names: Jones, James (James Ray), author. | Telionis, Demetri P., author.
Title: Aeroform : designing for wind and air movement / James Jones and
   Demetri Telionis.
Description: New York : Routledge, 2022. | Includes index.
Identifiers: LCCN 2022014401 | ISBN 9780367766184 (hardback) | ISBN
   9780367766191 (paperback) | ISBN 9781003167761 (ebook)
Subjects: LCSH: Wind resistant design. | Buildings—Aerodynamics.
Classification: LCC TA658.48 .J66 2022 | DDC 624.1/75—dc23/eng/20220722
LC record available at https://lccn.loc.gov/2022014401

ISBN: 978-0-367-76618-4 (hbk)
ISBN: 978-0-367-76619-1 (pbk)
ISBN: 978-1-003-16776-1 (ebk)

DOI: 10.4324/9781003167761

Typeset in Univers
by Apex CoVantage, LLC

Dedicated to Jared Jones

Contents

Figures

Figures

Figures

Tables

Preface

Aero: from the Greek word for air, defined by Webster as *a combining form meaning "air," used in formation of compound words*.

Form: *the essential nature of a thing as distinguished from its matter*. Thus, AEROFORM

Aeroform, as applied to architecture, is a proposition that the essential nature of a work of architecture can be informed by the influences of wind and airflow and that these influences can be expressive both performatively and aesthetically.

AEROFORM was a natural outcome of a collaborative relationship between the authors that began nearly 15 years ago. James (Jim) Jones is a designer, researcher, and professor of architecture whose focus has been on systems integration, environmental design, and resource conservation. The synthesis of these subjects led to his explorations into the opportunities for natural ventilation in buildings.

Demetri Telionis, before retiring, was a professor of engineering science and mechanics at Virginia Tech, with expertise in fluid mechanics, and he competitively races high-performance sailboats. In 2006, the two began working together on the design development, performance assessment, and patent of the V2T roof vent system. For this, Chuck and Pat Johnson, founders of Acrylife Inc., a roofing company from Wytheville, Virginia, brought to Dr. Jones an idea for a roof vent with two opposing conical tubes to take advantage of the Venturi effect to generate low pressure that can be transferred under roof membranes to counteract the uplifting forces created on roofs during high wind events. Dr. Jones saw the directional and time-responsive flaws in the proposed design and recommended a more omnidirectional geometry, such as two opposing pyramids, one above the other. Jones brought the idea to Dr. Telionis, which began the collaborative relationship and eventually led to the final dome shape for the V2T vent. The project evolved into the master of science thesis by Jones's student, Elizabeth Grant, and together the three used the aerodynamic wind tunnels at Virginia Tech to develop and optimize the design.

What is noteworthy is that when they would schedule to meet, Demetri would typically ask Jim, "Where should we meet, my office or yours?"; to which Jim would always answer, "Yours." You see, Demetri held the Frank Maher Professor position within the College of Engineering, a position that came with a premium office on the third floor of Norris Hall overlooking the Drillfield – the heart of Virginia Tech's campus. Demetri's office was large enough for not only his office furniture but also a seating area with a couch, comfortable chairs, and a coffee table, ideal for conversation and sketching ideas. In addition, the windows were operable and were typically cracked open just enough for natural ventilation. While sitting and talking about the design of the roof vent, we could feel the gentle wisps of fresh outdoor air pass over our bodies, providing a positive biophilic connection between indoors and out. Being educators in architecture and engineering, the two recognized the potential in merging their two perspectives on wind and airflow, particularly related to architectural design. AEROFORM is the result of this interdisciplinary synthesis, a merging of aerodynamics and architectural design.

AEROFORM is also a challenge to architects and architectural engineers to find the formative opportunities of wind and airflow on architectural form.

Acknowledgments

Many of the figures used in AEROFORM were developed by the authors. The digital preparation of these figures required many hours of effort. For this, the authors would like to thank Seyedreza Fiteminasab, Mansoureh Jalali, Sara Saghafi Moghaddam, Mitra Bagheri, and Ha Nuel Lee.

Chapter 1

Introduction

Wind: a natural movement of air of any velocity especially: the earth's air or the gas surrounding a planet in natural motion horizontally; an artificially produced movement of air; a force or agency that carries along or influences; a destructive force or influence. [1]

For thousands of years, the importance of the wind has been recognized. In the arts, literature, and in ancient religions the wind has been celebrated and worshipped. For the ancient Greeks, the Anemoi was the name given to the group of wind gods, such as Boreas, Zephyros, Notus, and Eurus who deified the differing directions of the wind. These gods served Aeolus, the only Greco-Roman god that could control the winds. When wind blows over narrow objects, such as cables, or slender columns, it excites the structure in specific frequency oscillations and generates musical tones known as Aeolian sounds.

Aeolian: giving forth or marked by a moaning or sighing sound or musical tone produced by or as if by the wind. [2]

Like many early belief systems, the Egyptians too had a group of gods that represented the winds in various forms. Referred to as "He who rises," Shu was the Egyptian god of peace, lions, air, and wind.

DOI: 10.4324/9781003167761-1

In poetry, words have been used to give visual impressions of the wind. Among the hundreds of poems written about the wind, John Coldwell wrote "The Shape of the Wind," in which he begins,

> The arctic blasts turn men into plaster casts
> They take the shape of the wind
> The ragged sails carved by the gales
> They take the shape of the wind
> . . .

The shape of the wind is a beautiful poetic presentation that at first may defy logic. But actually, the wind does have shape. In engineering terms, we can call its shape flow patterns or streamline systems. Surely, we cannot see the shape of the wind, but actually, we can. We can see the flow patterns of vortices generated behind a building corner by watching leaves recirculating near the ground. We can see some flow patterns betrayed by snowflakes. In engineering science, we can see the flow patterns by seeding air motion with small particles. This is called flow visualization and is a discipline within aerodynamics. Visualization techniques do not require sophisticated engineering instrumentation and can prove quite useful in architectural design, as discussed in this book.

As designers of built environments, architects should perhaps ask if architectural form might also take the shape of the wind. Architectural design is, in part, a process of responding to context, where contextural considerations include the site, culture, economic, societal, environmental, etc. Today, architects are being asked to design buildings that efficiently use resources, particularly energy to heat, cool, and light buildings. When resource conservation is a goal, the environmental context must inform design. This typically includes considering the climate and the daily and seasonal patterns of temperature and solar intensity. But the wind, too, is an important climatic factor that should influence architectural form. AEROFORM is both an introduction to the influences of the wind and air motion on architectural form and a challenge to designers to express those influences.

The historical significance of the wind on our lives is present not only in literature but can also be found in very early works of architecture. Built in the 2nd century BC, the Tower of the Winds, also known as the Clock of Andronicus Cyrrhestes, is a 44-ft (13.5 m) tall tower that tells time through a series of nine sundials (Figure 1.1). The tower originally had a bronze weathervane atop the conical roof to indicate the direction of the wind. Located on the eastern side of the Roman agora of Athens, the octagonal tower gets its name from the personification of the wind through carved reliefs on the top of each of the eight sides. [3] The representations are for each of the eight wind gods and the work was influential on future architects such as Vitruvius.

There are many names for winds derived from localities or from the squalls which sweep from rivers or down mountains. Marcus Vitruvius Pollio [4]

Completed in 2013, in the middle of the Tokyo Bay sits a more contemporary Tower of the Wind. "Kaze no to" in local language, this Tower of the Wind is two blue and white striped oval-shaped structures rising from a white circular base (Figure 1.2). Perhaps not as noble as the Clock of Andronicus Cyrrhestes, the Tokyo Tower of the Winds serves as

Figure 1.1 Tower of the winds, Athens Greece, source: Wikimedia

the ventilation air intakes and exhausts for the Tokyo Bay Aqualine, an undersea tunnel that runs 6 miles (9.6 km) at a depth of 130 ft (40 m) from Yokohama to Chiba. [5]

Yet another Tower of the Winds can be found in the city of Yokohama. A landmark in the center of a roundabout near the Yokohama train station, this 70-ft (21-m) tall tower also serves a ventilation purpose for the underground shopping center beneath it (Figure 1.3). Designed by Toyo Ito, the tower is an oval aluminum cylinder encasing synthetic mirrored plates. Lights positioned between these layers give the tower an appearance of a giant kaleidoscope. The tower has 1,280 small lights and 12 bright lights vertically arranged

(C) JUNKO AKISAWA

Figure 1.2 Tower of the Wind in the Tokyo Bay serves as a ventilation tower, source: Junko Akisawa

as neon light rings. Computer-controlled floodlights create patterns of light according to the time of day. The intensity of the floodlights varies with the wind, resulting in an environmentally responsive experience in architecture and light. [6]

For many professionals the understanding of wind and airflow is essential. Aerodynamic engineers, for example, must understand the effects of airflow over a wing and the resulting pressures that generate lift when designing aircraft. Similarly, the automotive industry is using wind tunnel testing and computational fluid dynamic (CFD) simulations to understand the interactions between wind, shape of the vehicle, and fuel economy. The Aptera 2e (Figure 1.4) is an electric vehicle aerodynamically designed for a drag coefficient of 0.15, about half that of a typical subcompact car. [7] Unfortunately for architects, too often the formative influence of wind on our buildings is ignored or not well understood, with only a technical footnote to address standards such as 2018 International Building Code and American Society of Civil Engineers (ASCE)/SEI 7–16. This book seeks to offer a remedy to this situation.

It could easily be argued that wind and air movement are essential for human existence. Wind plays a key role in the transfer of heat from the equator to the poles and for global temperature regulation. Whether regionally or locally, wind helps dissipate and move heat from hot to cool locations. Without the wind, Earth would experience temperature extremes far beyond those encountered during man's relatively short existence. Few architects formally embrace or acknowledge the importance of the wind in our lives or on our buildings.

Figure 1.3 Toyo Ito's Tower of the Winds, Yokohama, Japan, source: Wikipedia

In the book *Water and Architecture*, Charles Moore writes of the water cycle,

> Gusty winds push and pull the clouds around the atmosphere until the right conditions allow molecules to condense and fall back to the earth as rain, sleet, or snow to be absorbed into the terranean system, where the process (water cycle) can begin again. [8]

The wind and sun drive the water cycle and without the wind, there would not be life on earth as we know it.

An example of a regional wind-influenced condition that seems to make national news each year is the seasonal Santa Ana winds in the Los Angeles California basin

Figure 1.4 Aptera 2e low drag car, source: Wikipedia

(Figures 1.5 and 1.6). Named after Southern California's Santa Ana Canyon, the Santa Ana is a blustery dry and warm (or hot) wind that blows out of the desert. As described by Robert Fovell, a

> popular misconception that the winds are hot owing to their desert origin, actually the Santa Anas develop when the desert is cold, and are the most common during the cool season stretching from October through March. High pressure builds over the Great Basin (e.g. Nevada) and the air there begins to sink. However, this air is forced downslope which compresses and warms it at a rate of 29 °F per mile about (10 °C per kilometer) of decent. As the temperature rises, the relative humidity drops; the air starts out dry and ends up at sea level much drier still. The air picks up speed as it is channeled through the passes and canyons.

This is shown in Figure 1.5, while Figure 1.6 shows an annual wind wheel (rose) for Los Angeles. [9] These winds dry out vegetation, increasing the danger of wildfire; they create turbulence and wind shear and cause choppy surf conditions in the Southern California Bight. These winds often cause millions of dollars in fire damage in the Los Angeles area as experienced at a record pace in 2020. For those living in the Los Angeles valley, the Santa Ana winds both directly and indirectly impact their lives.

It is important to understand that there is feedback between the wind and an object, be it solid or flexible. When a flag is immersed in a steady, uniform wind, it does not remain straight, aligned with the direction of the wind, as shown in Figure 1.7. It flutters

Figure 1.5 Santa Ana wind map for Los Angeles, source: United States Geological Survey

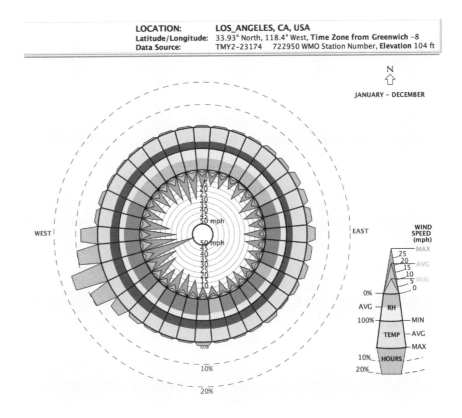

Figure 1.6 Annual wind rose for Los Angeles, source: author's image

Figure 1.7 Flag flying in the wind, source: Freepik

and forms elegant moving waves. It interacts with the wind, feeds back to the wind, and contributes to shedding air vortices downstream. When a steady wind flows over a solid body like a building or a bridge, it forms eddies, but this feedback induces unsteady loads on the structure. What is interesting is that the oncoming wind could be perfectly smooth and steady, and the structure has no moving parts. And yet vortices shed and move with frequencies that can be easily measured. The feedback between the wind and the structure gets one step further when these unsteady flow patterns actually excite the structure and induce vibrations that could be catastrophic. These phenomena are discussed in more detail in this book, and examples are presented of this type of catastrophe like the collapse of the Tacoma Narrows steel bridge.

A natural example of an aeroform, a form shaped by the wind, as proposed in this book, is the sand dune. *A hill or mound of sand shaped by the wind*, sand dunes are *dynamic structures migrating in the direction of effective, sand transporting winds, typically with gentle slopes on the windward side and steeper slopes on the leeward side* (Figure 1.8). *Dunes that move actively are called free dunes and reflect most dynamically the interaction between fluid atmospheric winds and moving sand.* [10]

Cresecentic dunes are crescent-shaped ridges of sand that form in response to a fairly unidirectional wind pattern. In Namibia, sand dunes can reach 800 ft (240 m) in height and are said to 'sing' due to the vibration of the sand grains driven by the wind. Often, sand dunes are picturesque examples of natural formations (aeroforms) produced by wind.

Another example of an aeroform found in deserts all over the globe is the yardang (Figure 1.9). The yardang is defined as *a streamlined protuberance carved from bedrock or*

Figure 1.8 African sand dunes, source: Wikipedia

Figure 1.9 Yardang in Meadow, Texas, source: Wikipedia

any consolidated or semi-consolidated material by the dual action of wind abrasion by dust and sand, and deflation which is the removal of loose material by wind turbulence. [11] *Yardangs become elongated features typically three or more times longer than wide and when viewed from above resemble the hull of a boat. Facing the wind is a steep, blunt face that gradually gets lower and narrower toward the lee end.* [12] *Yardangs are formed by* wind erosion, typically of an originally flat surface formed from areas of harder and softer material. The soft material is eroded and removed by the wind, and the harder material remains. The resulting pattern of yardangs is, therefore, a combination of the original rock distribution and the fluid mechanics of the airflow and resulting pattern of erosion. [13]

While air movement is important for sand dune and yardang formation, and global and regional thermal regulation, it is equally important at the human scale. Air movement over the body and the resulting convective heat transfer is often an important mechanism for bio-regulation and metabolic balance (Figure 1.10). The American Society of Heating, Refrigerating and Air-Conditioning Engineers (ASHRAE) Standard 55 – *Thermal Environmental Conditions for Human Occupancy* [14] indicates that for thermal comfort, the environmental factors along with air movement are air temperature, humidity, and surface temperature of nearby objects – which are the radiation sources and sinks that interact with the body. Increasing the airflow rate over the skin increases convective heat transfer and provides a desirable cooling effect in the summer or the undesirable condition of draft in winter.

Draft: The unwanted local cooling of the body caused by air movement. [15]

Beyond the measurable dimensions of airflow rate and temperature and their effects on thermal comfort, in her book *Thermal Delight in Architecture*, Lisa Heschong suggests that the presence of a summer breeze can provide a positive biophilia response where such stimuli might be counted with touch, taste, smell, sight, and hearing as one of the fundamental human senses. [16] Recognizing this, design strategies such as natural ventilation can not only reduce energy consumption but also can be linked to greater occupant satisfaction with indoor environments. This immeasurable dimension is demonstrated in the Bluewater Mall project in Dartford, United Kingdom, where the natural ventilation strategy was as much about bringing the condition of the city street, with its variation in airflow, to the contemporary mall concourse, as it was about reducing energy consumption. Figure 1.11 illustrates the interior of the Bluewater Mall with its circular ventilation openings above. The same may be said for Mario Botta's chapel for Santa Maria degli Angeli Monte Tamaro (Figure 1.12) near Bellinzona Switzerland, where the regularly spaced openings of

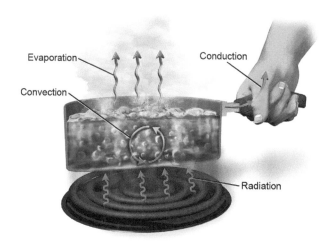

Figure 1.10 Thermal exchange mechanisms, source: Wikipedia

Figure 1.11 Bluewater Mall interior, source: Wikipedia

Figure 1.12 Alternating openings in Monte Tamaro chapel, source: WordPress

the enclosed approach to the chapel not only regulate light but also give rhythm to the wind as one passes along the path while feeling the rise and fall of the breeze. Here airflow and architecture make present the dialogue between indoors and out while contributing to one's self-awareness, to a perceptual condition of orientation and identity as promoted by architectural theorist Christian Norborg-Schultz.

The implication from ASHRAE Standard 55 and projects such as Bluewater and the Monte Tamaro chapel is that the influence of wind and air movement can serve as a bridge between the measurable and immeasurable while offering an opportunity for expression.

Architecture should be about people and is, in part, concerned with making "places." Vitruvius suggests

> the architect should know those inclinations of the heavens which the Greeks call climates, and know about airs, and about which places are healthful and which disease ridden, . . . for without these studies no dwelling can possibly be healthful. [17]

Beyond Vitruvius's somewhat analytical assessment of "places," we must acknowledge that because architecture is often concerned with the making of "places," the influence of wind and air motion on the place must be part of the architect's understanding.

A concept that designers should understand is the harnessing of the wind. The term of course comes from the harnessing of horses. But harnessing does not only mean that you can hold and restrain the animal. Harnessing implies that you can direct the great power of a horse to a certain purpose. You can ride it or harness it to a cart to carry heavy loads. The same holds for harnessing the wind. Not only can we restrain and control the power of the wind to prevent it from damaging our structure but also redirect the power of the wind to our benefit. A typical and well-known example is harnessing the power of a hurricane to hold a roof down and prevent it from being lifted off. A roof is lifted if the low pressure generated on the roof by the wind is much lower than the inside pressure. Windows open or even cracked open on the windward side of the building allow high pressure to be established inside the building. As a result, very low pressure on the roof and very high pressure from the inside lift the roof. But if windows are closed on the windward side and open on the lee side where the pressure is very low, then very low pressure is established inside the building, and it sucks the roof down.

It must have been exciting for researchers and designers when they realized that they can actually guide the harnessing toward desired outcomes. For example, in chemistry, a catalyst in minute amounts would guide the direction of chemical reaction. In aerodynamics, a very small disturbance of the flow over a wing can trigger the development of large turbulences downstream. As a result, the large-scale flow over the wing rearranges the pressure distributions and reduces drastically the drag of the wing. The authors of this book, in collaboration with their students, have invented and patented the V2T roof vent (Figure 1.13), described in Chapter 5, to generate very low pressure under the roofing material and thus harness the power of a strong wind to suck down the material on the roof and prevent it from tearing. Examples of harnessing the wind to our advantage we can find in sailboats. It is beautiful to see how a boat can actually sail in a direction opposite to the wind, in nautical terminology a boat is beating windward. It is also exciting to see modern sailboats sailing at speeds even higher than the wind speed.

The city of Amsterdam in the Netherlands has become a *place* for international tourism in large part due to the harnessing of the wind by windmills to pump water from low-lying areas while also using them for milling grain and other industrial processes that contributed to the development of the city. As a result, the windmill has become an iconic symbol for the city (Figure 1.14). While only about eight functioning windmills still exist in

Figure 1.13 V2T roof vent system (courtesy V2T Roof Systems Inc.)

Amsterdam today, arguably, Amsterdam, as a "place," would not exist today if not for the harnessing of the wind.

Understanding and utilizing the wind is not an exclusively modern condition. On the contrary, it could be argued that man's utilization of wind is much less prevalent today when compared to the past. In ancient times, boats and ships were rigged with sails to take advantage of wind forces to propel the vessel through the water for exploration, transportation, and to gain a competitive advantage in warfare. As far back as 8,000 years ago, Cucuteni-Tryptillian sailors in Eastern Europe were harnessing the wind and developing sail technology. By 3000 BC, proto-Austronesian people were sailing across the islands of the South Pacific. Today, the America's Cup continues this nautical tradition of harnessing wind power for competitive yachting.

Sailing affords a wonderful opportunity to study the shape of moving air through its effect on sails and the water, Figure 1.15. In a sense, sailing is the best way to simulate the effect of tall structures on the airflow around and among them. And just as sailboats can use the wind to achieve different points of sail, buildings can use their outer skin or layers of their structure to harness the wind for their purpose. When we look at sailing a little more closely, we can also observe how finely we can harness the power of the wind to affect changes in velocity, heeling moment, and lift. The sail controls on a high-performance boat's mainsail, for example, allow near infinite adjustment of depth of the sail but also the location

Figure 1.14 De Gooyer windmill Amsterdam, source: Wikipedia

Figure 1.15 Inland lake sailing, source: author's photo

of maximum or minimum depth, as well as the location of the center of effort depending on the boat, but pretty much expressible in general terms, the amounts of change needed in the trim of sails to achieve balance are minute compared to the total sail area of the sail plan. [18] While buildings are static constructs (one would hope) the forces acting on them are the same as on a mobile entity such as a sailboat. It stands to reason that careful study of kinetic or static surfaces can greatly affect design issues such as the energy needed to supplement natural ventilation. A proper response to the wind will not only give the boat the desired motion but also make the boat look aesthetically pleasing, while a bad response to the wind will not only make the boat have to fight the wind but also creates a rather disturbing visual effect. Through this book, perhaps a similar approach could be applied to architecture.

Another example of the power of wind and high-speed aero design is the story of Rosemeyer's fatal accident. "*Herr Berndt Rosemeyer, the brilliant racing driver employed by the Auto-Union company, was killed this morning on the Frankfurt-Darmstadt Autobahn, while attempting to beat speed records set up earlier in the morning by Herr Carraciola of Mercedes-Benz [19]. When he was driving at over 400 kilometers per hour, 250 miles an hour, a gust of wind carried the car to the extreme outer edge of the track. Rosemeyer was unable to bring it again into the center; it collided with a stone pillar of a bridge carrying an ordinary road over the Autobahn, and he was thrown out of the car and killed.*" We can only hope that such catastrophic results from wind do not impact our buildings.

Another historical example of wind utilization can be found in agrarian societies where windmills were often used to elevate water from wells for drinking, irrigation, and watering livestock. Today, wind turbines generate electricity and represent a growing alternative to fossil fuel power sources (Figure 1.16). Applications such as these suggest that we

Figure 1.16 Wind turbines for electrical generation, source: Wikipedia

may be returning to a condition where the wind can play a larger role in our buildings and infrastructure. Designers of built environments should have a fundamental understanding of wind and air movement as they design the buildings that surround our lives.

Some architects, albeit rather few, have embraced the wind as a design element.

> I seek to instill the presence of nature within an architecture austerely constructed by means of transparent logic. The elements of nature – water, wind, light and sky – bring architecture derived from ideological thought down to the ground level of reality and awaken man-made life within it. [20]

Through this quote, Tadao Ando suggests that architects need to understand the influence of the elements of nature, including wind, on the man-architecture dialogue. Ando hints that wind can become an informant to architectural design and that architecture should be of its place, site, and climate, including wind conditions. This can be reflected in his serene design for the Church of the Wind located atop Mt. Rokko near Kobe Japan (Figure 1.17).

Related to a position of Critical Regionalism as put forth by Kenneth Frampton, suggestions such as Ando's to weave the building into the local environment and culture are not new. For example, the hemispherical igloo structures of the American Eskimos (Figure 1.18) not only minimize the surface-to-volume ratio but also present minimal exposure to wind and infiltration, which is most appropriate for harsh cold climates. Similarly, site strategies such as earth berming (partial submersion of the building into the ground) take advantage of topography and sloping terrain to partially submerge and protect the building from winter winds. Frank Lloyd Wright's hemicycle concept as applied in the Jacobs House is an example of this design response to wind, sun, and site in the Midwest United States (Figure 1.19). Such telluric strategies can express the relation of the building to Earth for the purpose of responding to the wind.

In hot climates, the historical response was not one of protection from the winds but one of capturing and utilizing the prevailing breezes for convective cooling and natural ventilation. The wind towers and malquafs of the Middle East (Figure 1.20), and the courtyard layouts of North Africa and US desert southwest demonstrate indigenous strategies to understand and harness wind and airflow forces toward improving the human condition. Similarly, the tipi tent structures of Native Americans both served their nomadic lifestyle and supported natural air movement from the low perimeter upward to an opening at the apex of the structure. This was efficient for both cooling and for removal of smoke from fires located near the center of the tipi. Unfortunately, it seems that some of these lessons have been lost with the increased dependence on mechanical ventilation.

Today, architects are being asked to design "sustainable" buildings with low energy consumption and high indoor environmental quality. Strategies such as natural ventilation that harness the wind are being considered to meet these objectives. In her dissertation titled *A Decision-Support Framework for Design of Non-residential Net-Zero Energy Buildings*, Railesha Tiwari [21] found that natural ventilation and daylighting were among the most common design strategies for achieving net-zero. Effective design for natural ventilation must begin with understanding wind effects on buildings.

When designing buildings for natural ventilation, simulation techniques such as wind tunnel models and CFD are more frequently being used to support decision-making.

Figure 1.17 Church of the wind by Tadao Ando, source: Wikipedia

CFD studies have become fairly common presentations at architectural conferences. Unfortunately, many of these studies indicate a lack of understanding of the underlying fluid flow principles, improper assumptions for wind direction and speed, and misinformed boundary-layer profiles that can significantly influence the reliability of the results. For wind tunnel modeling, scaling factors related to the Reynolds number must be well understood if the results are to relate to the as-built conditions. This book seeks to build this understanding for those using these analysis techniques.

For these and other reasons the architect should be well-versed in principles and issues of wind and air movement in and around buildings. AEROFORM is a synthesis of

Figure 1.18 Eskimo igloo, source: author's figure

Figure 1.19 Frank Lloyd Wright Jacobs House, source: Archipedia

Figure 1.20 Malquaf wind tower, source: Wikipedia

architectural design and engineering principles related to aerodynamics. Through this synthesis, AEROFORM demonstrates that wind and air movement can be elegant formative influences on architectural design. The understanding of aerodynamic principles can be applied to architectural design strategies that reduce wind-driven pressures on building enclosures, thus decreasing infiltration and energy consumption to create pressure differences that promote natural ventilation and a strong connection between inside and out or reduce the impacts from high wind events such as hurricanes. AEROFORM prepares the architect or architectural engineer to design buildings that are visually expressive of a dialogue between wind and built form.

Beyond this introduction, Chapter 2 presents the fundamental concepts in aerodynamics that will be referenced throughout the book. This is intended to establish a foundation of understanding in the reader. The concepts and principles are described in text and images to promote understanding of visually oriented learners, such as students of architecture. The chapter starts with fundamental concepts such as materials, forces, equilibrium, pressure, and hydrostatics. The chapter proceeds with a discussion of fluid flow patterns and their representation. The authors see the content of this chapter as a resource for both designers and engineers.

Chapter 3 introduces the fundamentals of wind and airflow. This begins with the global scale and large convective cells that result in the trade winds, moves to coastal wind patterns, and then on to local influences of landforms and topography on air movement. This chapter references concepts introduced in Chapter 2 and builds on these concepts with more in-depth description of fluid flow. This also includes descriptions of phenomena, such as the boundary layer that develops over Earth's surface in response to different terrain conditions. The chapter connects these concepts to pressure that develops on the enclosure of buildings.

Chapter 4 describes design strategies than can either reduce or take advantage of wind-driven pressures on buildings. The chapter begins by presenting how wind and buildings interact and the resulting pressure distributions that develop on walls and roofs. It presents examples of design strategies that take advantage of these interactions. The chapter begins with general wind-driven pressures on buildings and translates those pressures into design responses such as siting the building, use of vegetation, building geometry, and wind screens.

Chapter 5 addresses the specific issue of high wind events. The chapter introduces issues related to the roof construction and geometry for reducing wind pressures and uplift during high wind events such as hurricanes. Among the topics introduced are the authors' recent research on this topic and the invention of a roof venting system that relieves some of the uplifting pressures that tend to detach roofing membranes on low-sloped roofs during high wind events. The chapter concludes with design considerations for straps and tie-downs that are required in high wind zones.

Chapter 6 addresses the popular topic of natural ventilation. Natural ventilation is a common design strategy when low energy consumption and sustainability are design objectives. This chapter presents concepts for developing pressure differences that drive natural ventilation through wind and thermal buoyancy. Different approaches to natural ventilation are presented through built examples. This includes single-sided and cross-ventilation. Ventilation and concepts such as the neutral pressure level. The chapter also introduces easy-to-understand tactics and rules of thumb for application.

Chapter 7 discusses concepts and issues related to forced ventilation in buildings. This includes principles of air pressurization from fans, air distribution in ducts, air diffusion into rooms, and different patterns of room air movement. This introduces strategies such as displacement and piston-flow ventilation that likely will grow in popularity due to COVID-19.

Chapter 8 presents issues related to the measurement of air movement, including instrumentation for monitoring wind and airflow. The chapter discusses experimental modeling of airflow in and around buildings and issues related to scaling in a wind tunnel, and the relation to the Reynolds number. This is particularly relevant for teaching issues of wind impacts on buildings using wind tunnel simulation, where scaling factors are very important. This chapter also addresses the use of simulation tools such as CFD.

References

[1] Merriam-Webster (2022) *Definition: Wind*. Available from: www.merriam-webster.com/dictionary/wind [Accessed September 21, 2019].

[2] Merriam-Webster (2022) *Definition: Aeolian*. Available from: www.merriam-webster.com/dictionary/aeolian [Accessed September 21, 2019].

[3] World History Encyclopedia (2022) *Tower of the Winds*. Available from: www.ancient.eu/article/1044/tower-of-the-winds/ [Accessed September 21, 2019].

[4] AZ Quotes (2022) *Marcus Vitruvius Pollio Quotes*. Available from: www.azquotes.com/author/38395-Marcus_Vitruvius_Pollio?p=3 [Accessed September 21, 2019].

[5] Atlas Obscura (2022) *Tower of Wind, Kawasaki, Japan*. Available from: www.atlasobscura.com/places/tower-of-wind [Accessed September 22, 2019].

[6] Archute (2022) *Toyo Ito's Tower of Winds, a Technological Sculpture in Yokohama Japan*. Available from: www.archute.com/toyo-itos-tower-winds-technological-sculpture-yokohama-japan/ [Accessed September 22, 2019].

[7] Aptera (2022) *Aptera Solar Electric Vehicle*. Available from: www.aptera.com [Accessed September 22, 2019].

[8] Moore, C. and J. Lidz (1994) *Water and Architecture*. New York, NY, Harry N. Abrams, Inc. Publishers.

[9] Fovell, R. (2002) *The Santa Ana Winds*. Available from: www.atmos.ucla.edu/~fovell/ASother/mm5/SantaAna/winds.html [Accessed September 22, 2019].

[10] Society (2022) *Dune*. Available from: www.nationalgeographic.org/encyclopedia/dune/ [Accessed September 22, 2019].

[11] Wikipedia (2010) *Desert Processes Working Group: Yardangs, at the Wayback Machine*. Available from: https://en.wikipedia.org/wiki/Yardang [Accessed September 22, 2019].

[12] Open University Geological Society Mainland Europe Branch (2022) *Yardangs*. Available from: Martian Fleets – August 2003 [Accessed September 22, 2019].

[13] Wikipedia (2010) *Desert Processes Working Group: Yardangs, at the Wayback Machine*. Available from: https://en.wikipedia.org/wiki/Yardang [Accessed September 22, 2019].

[14] ASHRAE (1992) *Thermal Environmental Conditions for Human Occupancy – Standard 55*. Atlanta, GA, The American Society of Heating, Refrigerating and Air-Conditioning Engineers, p. 2.

[15] ASHRAE (1992) *Thermal Environmental Conditions for Human Occupancy – Standard 55*. Atlanta, GA, The American Society of Heating, Refrigerating and Air-Conditioning Engineers, p. 3.

[16] Heschong, L. (1999) *Thermal Delight in Architecture*. Cambridge, MA, The MIT Press, p. 18.

[17] AZ Quotes (2022) *Marcus Vitruvius Pollio Quotes*. Available from: www.azquotes.com/author/38395-Marcus_Vitruvius_Pollio?p=3 [Accessed September 21, 2019].

[18] Rott, H. (2016) *Personal Reflection on Competitive Lake Sailing*. Blacksburg, VA, Department of Biological Systems Engineering, Virginia Tech.

[19] Wikipedia (2022) *Bernd Rosemeyer*. Available from: en.wikipedia.org/wiki/Bernd_Rosemeyer [Accessed September 23, 2019].

[20] Dal Co, F. (1995) *Tadao Ando: Complete Works*. London, Phaidon Press, pp. 456–457.

[21] Tiwari, R. (2015) *A Decision-Support Framework for Design of Non-Residential Net-Zero Energy Buildings*. Ph.D. dissertation, Virginia Polytechnic Institute and State University, Blacksburg, VA.

Poem Citation

Poem Hunter (2009) *Shape of the Wind by John Coldwell*. Available from: www.poemhunter.com/poem/the-shape-of-the-wind/ [Accessed September 23, 2019].

Chapter 2

Basic Concepts of Fluid Mechanics

Our world is immersed in fluids. Most of us are not always aware of the existence of the "fluid" we cannot escape from, air, unless we feel a light breeze or, much more forcefully, if we find ourselves in the middle of a storm. The motion of this fluid is then strongly felt, and it could be catastrophic. The other fluid we sometimes literally swim in of course is water. There are many other fluids, some natural and others man-made, but in this book, we will concentrate on the study of the behavior of air while occasionally using water flow for illustration.

Fluid mechanics is a discipline in physics and in engineering mechanics dedicated to the study of the motion of fluids and their effect on solid bodies. The behavior of most fluids is governed by the same laws. In other words, we can study the behavior of air or water using the same analytical and experimental tools. The motion of fluids is intriguing, but what we really need to know in engineering and architecture is their effect on solid bodies such as buildings. Wind blowing over a structure develops forces that vary with the wind direction and magnitude. These forces could damage objects like buildings or other standing structures; they could overturn vehicles or ships. But we have also harnessed forces generated by moving fluids to lift vehicles like airplanes or helicopters, to extract energy via wind turbines, or to control and condition our environment. Understanding and controlling these forces requires the study of fluid motion and the forces that are developed on solid surfaces a fluid is moving over. This book is devoted to presenting the basic concepts of fluid motion and its effects on built environments.

For well over 100 years, fluid mechanics research has been conducted using analytical, experimental, and numerical methods. Most of these methods are advanced, and only specialists can make contributions using them. However, today there is a wealth of

DOI: 10.4324/9781003167761-2

information that practitioners of engineering and architecture can use to design machines and structures with no special training in the use of advanced tools. Typical examples are computer codes that solve fluid mechanics problems. Such codes are the result of numerous advanced studies. But now these can be provided in the form of computational fluid mechanical packages (CFD) that are commercially available. Without knowing the detailed function of the algorithms incorporated in the software, a user can employ CFD to calculate the forces generated over a building or a machine component by entering the shape and size of the design and then allowing the software to generate the desired information. However, as with any analytic tool for supporting design decision-making, the tool should not be treated as a black box. When using CFD, the reliability of the results will depend on the quality of the input, and here the user should be familiar with the fundamentals of fluid mechanics.

The study of the material in this book, or taking a course based on it, will not enable a student to carry out detailed quantitative estimates of fluid motions or loads generated by fluids. The aim here is to provide understanding of the basic principles of fluid mechanics; familiarize the readers with the terminology of fluids engineering, which will allow them to interact directly with the specialists; and facilitate their understanding of the results produced by powerful numerical or experimental methods. The material presented here will hopefully be useful to architects or engineers specializing in disciplines other than fluids engineering. It will thus allow them to understand the significance of fluids engineering results and communicate more efficiently with the specialists. The student of this book may even be able to employ a commercially available CFD code or some simple experimental tools to estimate flow properties or loads on a structure. These capabilities could then serve as guidance in the original design or a modification of an existing design.

It is assumed that the readership has some basic understanding of mechanics. But some very basic mechanics terms and principles are included here.

Materials

Our world is filled with **materials**. These are the things that we can touch, lift, mold and experience in many different ways. It is what we see around us. There are natural materials, which are what we find in nature, and synthetic materials, which are the materials we formulate for special use. Wood is a natural material used in construction. Iron is a natural material, but it requires special processing to produce steel with the special properties required in practice. Synthetic materials like fiberglass and other composite materials with properties superior to natural materials are becoming more common. And there are of course biological materials created by living organisms.

In mechanics, we call **solids** the materials whose shape does not change if pushed, compressed, or otherwise contacted. As with most building materials, wood, iron, brick, and rock are solids. Actually, under certain conditions, the shape and size of all solids can be modified somewhat. Metals, for example, can be extended if loaded, but their extension is barely visible.

Fluids are materials whose shape is adjusted according to their environment. Water, for example, takes exactly the shape of a pot into which it is poured. There are two

Figure 2.1 Liquids take the shape of a container, source: Wikimedia

basic categories of fluids, namely **liquids** and **gases**. Liquids can adjust in shape, but their volume changes very little. If you pour 4 ounces of water into an 8-ounce glass, it will reach only halfway up to its rim. (Figure 2.1) Again, there are natural fluids like water, products of processed fluids like gasoline and diesel fuel, synthetic fluids like adhesives, and, finally, biological fluids like blood, milk, and many others. **Gases**, on the other hand, will expand and take the entire volume of a container they are forced into.

Material Properties

Materials are characterized by the compactness with which their weight is packed within the space they occupy. We measure the packing of the weight in a material with the quantity of **specific weigh**t, which is the weight of a unit volume of the material. The specific weight is usually symbolized by the Greek letter gamma (γ) and is measured in pounds per cubic inches or in pounds per cubic feet (lb/in^3 or lb/ft^3) in English units or in Newtons per cubic meter in SI units (N/m^3). A similar, and perhaps more common, measure of how compact is a material is the **density**, which is the mass of the material per unit volume. The physical quantity of **mass** is difficult to define in simple terms and will be omitted here. Under standard conditions, the specific weight of air is 0.0764 lb/ft^3 and of water is 62.3 lb/ft^3 or 0.036 lb/in^3 (at 70°F). Most people find it strange to think of the weight of air. The concept will become clear when we discuss hydrostatics and the effect of the weight of the air in the atmosphere.

Forces

Buildings experience many forces; the weight of a building exerts a downward force on the ground due to gravity. The wind both pushes and pulls on the walls of the structure, and at high velocities, it exerts uplifting forces on the roof that can be detached if not properly anchored. In mechanics, a **force** is defined as the action of one body on another. This is the action that could result in some deformation of a body to which the force is applied, or it could result in a change of its motion, as was the case of the wind pushing Rosenmeyer's car in Chapter 1.

If a body is at rest, a force exerted on it could get it moving with increasing speed, or if it is already in motion, the force can change its speed. A measure of the change of speed is called **acceleration**, as experienced when one pushes harder with the foot on the accelerator pedal of a car to increase speed. If Body A exerts a force on Body B, then Body B exerts an equal and opposite force on Body A. This is Newton's principle of action and reaction. For example, if a ladder, here body B, is leaning against the wall, Body A, as shown in Figure 2.2a, then the wall is exerting a force against the top of the ladder. This is indicated with the arrows in Figure 2.2b. This is the force that prevents the ladder from falling over. The ladder, Body B, is exerting an equal and opposite force to the wall, Body A, as shown in Figure 2.2c. A force could be concentrated – that is, exerted at a point. But forces could also be distributed over an area or even over a volume, as discussed later in this chapter.

When describing materials, physical quantities can be characterized in terms of their magnitude. A physical quantity that can be adequately described with just its numerical value is called a **scalar**. Temperature is a scalar quantity. It is enough to state that the temperature in a room is say 70°. It makes no sense to say that it is 70° upward, or to the right. But there are other physical quantities that require more than a numerical value to be described. In mechanics quantities that require both the magnitude and direction to be described are called **vectors**. Wind has both magnitude (speed) and direction and is, therefore, a vector quantity. Forces are vectors. Consider a weightlifter lifting weight (Figure 2.3). The weight is exerting a downward force that is balanced (the weightlifter hopes) by the upward lifting force of the weightlifter's strength. Similarly, a book resting on a table exerts a downward force that is counteracted by the uplifting force of the table (Figure 2.4). Vectors are usually depicted schematically with arrows, whereby the length of the arrow can represent its magnitude, and its direction can represent the direction of the force. We represent schematically the action of the book on the table by an arrow pointing downward, as shown

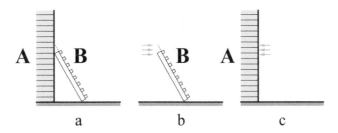

a   b   c

Figure 2.2 A ladder leaning against a wall, source: author's image

Figure 2.3 Counter forces of lifting weights, source: Wikimedia

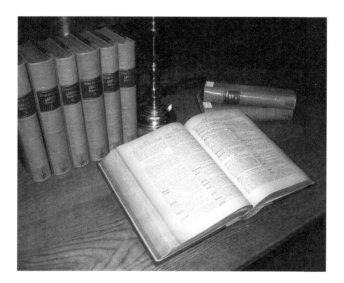

Figure 2.4 Book on a table, source: author's image

in Figure 2.5. According to the principle of action and reaction, the table is exerting a force on the book upward, as indicated in Figure 2.5. This is the force that prevents the book from falling downward. Buildings react to many forces in this way.

Similarly, water contained in a swimming pool is exerting a force on the retaining walls. In Figure 2.6, we show schematically the effect of the water on one of the retaining

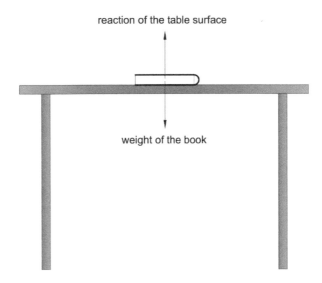

reaction of the table surface

weight of the book

Figure 2.5 Vector force representation, source: author's figure

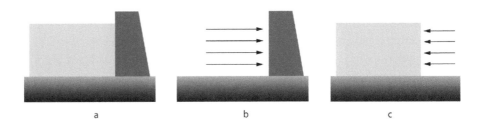

a   b   c

Figure 2.6 A wall retaining a body of water, source: author's figure

walls. This force is in a horizontal direction and can be denoted schematically by arrows, as shown in Figure 2.6b, where we "removed" the water, and replaced it with its action indicated by the arrows. The wall in turn is exerting forces on the water, which keep it from flowing away, as shown in Figure 2.6c. This is another example of action and reaction. Forces exerted by fluids are always distributed forces. Similarly, buildings are immersed in the air of the atmosphere, and the enclosure experiences distributed forces. These are distributed forces and can be represented by a number of small force arrows, as depicted in Figure 2.6. We will explain later how these forces are distributed over the area of contact between the wall and the water or air.

Similarly, wind exerts a force on a building enclosure, as in Figure 2.7. We will show later that the distributed forces exerted on the windward side of the enclosure are compressive, while on the roof and the leeward side, the forces generate suction. These are the net forces combining the effect of the external wind and the internal pressure and are shown in the figure as positive for compression and negative for suction.

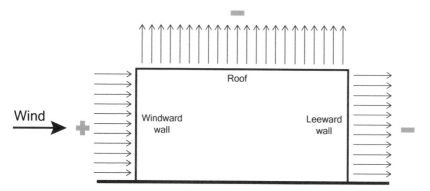

Figure 2.7 Wind forces on a building, source: author's figure

Gravitational Forces and Equilibrium

So far, we have seen forces exerted on a body by other bodies. There is another kind of force that all bodies experience. This is the force of gravity. The gravity force is usually referred to as the **weight** of a body. The gravitational force is exerted on the body by the earth, and of course, it is always downward. A gravitational force varies in relation to the size (mass) of the attracting body. For example, because the size of the moon is less than that of Earth, the gravitational force on the moon is smaller, and one would be able to jump quite high on the moon. On Jupiter, on the other hand, because of the greater mass than Earth, it would be difficult to walk, much less jump.

These concepts are discussed later in the book. Of interest for us here is the concept of **equilibrium**. Newton's first principle states that if all forces exerted on a body "balance each other out," then the body is at rest, or if the body was moving, it will retain its motion in the same direction and with the same velocity. The concepts mentioned in this paragraph are presented in very simple terms and will not be necessary for the presentation of fluid mechanics concepts. The statement that "forces balance each other out" in mechanics represents an involved vector operation. In its most simple terms, we can state here that forces exerted on the book of Figure 2.5 – namely, its weight – balances the force that the table exerts on the book. The book then is in equilibrium.

There are three forces exerted on the retaining wall of Figure 2.6, the force exerted by the water, the force exerted by the ground, and the gravity force. These are shown schematically in Figure 2.8.

Here F_w is the force exerted by the water on the wall; P_x and P_y are the horizontal and vertical components of the force exerted by the ground on the wall. P_y is the force exerted by the ground that carries the weight of the wall. P_x is the friction force that prevents the wall from sliding. W is the weight of the wall shown exerted at its center of gravity. To show how forces can actually "balance each other out," we will need mathematical tools that are beyond the scope of this book. But in this special case where all forces are aligned with the system of coordinates x, y, equilibrium dictates that $P_w = P_x$ and $W = P_y$.

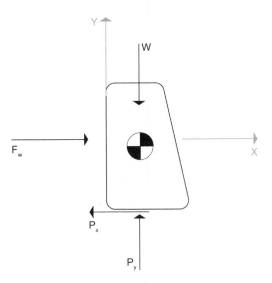

Figure 2.8 Forces on retaining wall, source: author's figure

In mechanics, a diagram of a body showing all the forces exerted to it by other bodies around it is called a **free-body diagram**. The diagram in Figure 2.8 is a free-body diagram of the retaining wall.

Hydrostatics

Fluids, such as air, can exert only distributed forces. Fluid forces are exerted on solid bodies regardless of whether the fluid is in motion or at rest. The force exerted by the water on the retaining wall of Figure 2.6 is distributed over its entire area. A numerical example, which will help us understand the concept of distributed forces is appropriate here. Consider a cubic foot container filled with water. The container dimensions are 12 inches width by 12 inches length and 12 inches height (Figure 2.9). One cubic foot of water weighs 62.4 lb (at 30°F). With base dimensions of 12 inches by 12 inches, the area of the base is 144 square inches. Thus every square inch of the base of this vessel is carrying (62.4 lb)/(144 in^2) = 0.433 lb/in^2. We often use psi to denote pounds per square inch. This value, 0.433 psi, is the pressure exerted per foot of height of a water column. In other words, a column of water 10 ft high will exert a pressure at the base of 10 X 0.433 = 4.33 psi. This is important for plumbing systems for pressure distribution and pump sizing, as a vertical pipe standing 30 ft high would have a base water pressure of about 30 X 0.433 = 13 psi.

In fluid mechanics, a force distributed over an area is called **static pressure**. In this numerical example, the pressure is measured in pounds per square inch, lb/in^2. The water exerts a pressure of 0.433 lb/in^2 on the bottom of the vessel per foot height.

Fluids exert pressure on solids, although not always equally. Any person diving into a pool of water experiences these pressures on his/her eardrums. The same may be experienced when taking off or landing an airplane or ascending or descending a hill in a

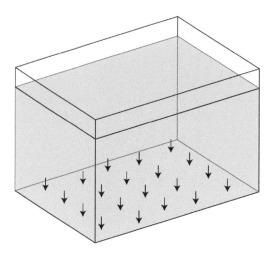

Figure 2.9 A vessel containing water, source: author's figure

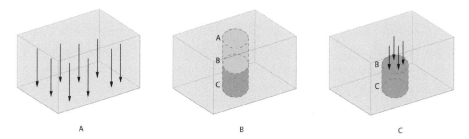

A   B   C

Figure 2.10 Weight transferred from one fluid element to another, source: author's figure

car. But fluid elements also exert pressure on neighboring fluid elements. In the example of Figure 2.9, elements of the fluid exert pressure on the bottom of the container but also on their neighboring fluid elements. Consider, for example, in Figure 2.10 the column AB, colored in orange just to help us visualize it, which sits on top of the column BC (colored in green). With a height of the orange column of say 3 inches and an area of 1 in², column AB has a volume of $V = 3$ in³. In this book, we will use a capital V to denote velocity and a script V – i.e., V to denote volume. We can now calculate the weight of this column, W using the specific weight of water.

$$W = \gamma V = (0.036 \text{ lb/in}^3) \bullet 3\text{in}^3 = 0.108\text{lb} \tag{2.1}$$

Thus, a force of 0.108 lb is exerted over the area of 3 in² of the green column, as shown schematically in Figure 2.10c, which corresponds to a pressure of

$$p = (0.108 \text{ lb})/(3 \text{ in}^2) = 0.036\text{psi} \tag{2.2}$$

This example nicely indicates that the pressure increases with depth because for a height of column AB of 6 inches, the force transmitted to the column BC will be 0.216 lb, and the pressure will be 0.072 psi. Again, we experience this in a pool as we become more aware of the water pressure on our eardrums as we dive deeper. In fact, it can be easily shown mathematically that the pressure increases linearly with depth. If we measure the depth from the surface of the water downward with the coordinate z, then the pressure at a depth z is given by

$$p = p_a + \gamma\,z \tag{2.3}$$

Here p_a is the value of the atmospheric pressure, which is the pressure at the free surface – i.e., at z = 0. Pressure is an internal feature of the fluid – i.e., a fluid element exerts pressure on its neighbor beneath it but also on its neighbor above it or, for that matter, its neighbor on its side. In fact, a liquid is pushing against the wall of the container with a pressure given by Equation (2.3).

Just like the columns of water in the domain of Figure 2.10 generate static pressure on the bottom of the container, a column of a few thousand feet of air in the atmosphere generates on the surface of the earth a significant static pressure. Most of us find it hard to conceptualize the "weight of air." But close to the seashore, the pressure generated by the weight of the air in the atmosphere is approximately 14.7 psi. This literally means that every square inch of our skin is pushed inward by 14.7 lb. The atmospheric pressure varies with elevation, with temperature, and with atmospheric conditions.

Students usually marvel at the fact that a 6-foot-tall person lying on the beach carries about 16,934 lb on his or her body; yes, about 17,000 lb is the weight of the atmosphere directly above his/her body. For this example, we made the simplifying assumption that the person has an average "width" of 16 inches and therefore an overall average area exposed to the atmospheric column above him/her of 1,152 square inches. Pressure is actually uniformly distributed all around his/her body, pushing inward with the same pressure. Without this pressure, a human being would literally explode. To allow astronauts to exit outside the pressurized cabin of a spacecraft where there is no atmospheric pressure, space uniforms are needed to generate large pressure on the astronaut's body. It is convenient for engineers and architects to run their calculations in terms of pressure values above the ambient atmospheric pressure. The difference between a pressure at a point p_s and the atmospheric pressure p_a is called gauge pressure, p:

$$p = p_s - p_a \tag{2.4}$$

With the specific weight of water about one thousand times higher than that of air, it only takes a water depth of 33 ft (10 m) for the hydrostatic pressure to double. Water pressure exerted to the walls of a container, like a tank, or a swimming pool, or the walls of a dam, generates significant static loads that should be accounted for by engineers and architects. Consider, for example, a wall containing water of a depth of 8 ft, as shown in Figure 2.11. The linear pressure variation of Equation (2.3) is presented in the figure with a linear profile of arrows that increase with depth. The pressure on the other side of the wall is the atmospheric pressure, and this too is denoted in Figure 2.11b with a row of equal arrows.

$$p = p_s - p_a$$

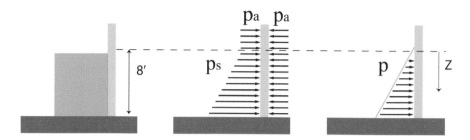

Figure 2.11 Hydrostatic pressure exerted on a retaining wall, source: author's figure

It is common practice to present static pressure data and results in terms of gauge pressure. This is because the atmospheric pressure varies very little with elevation, and for all practical purposes, we can assume that for heights of a few hundred feet, the atmospheric pressure does not change. The situation presented in Figure 2.11a then is depicted in terms of gauge pressure as shown in Figure 2.11c and is given by Equation (2.5):

$$p = p_s - p_a = \gamma z \qquad\qquad (2.5)$$

The maximum pressure is thus found at the maximum depth of 8 ft or 96 inches and is $p = \gamma z = (0.036 \text{ lb/in}^3) \cdot 95 \text{ inches} = 3.456 \text{ psi}$.

As discussed earlier, pressure is a scalar. But the effect of pressure distributions on surfaces gives rise to forces. One can visualize this effect by thinking that each one of the arrows representing pressure magnitude in Figure 2.11 is an individual force acting perpendicular to the surface. The combined effect for this triangular distribution of pressures is a force normal to the wall and equal to the surface of the triangle per unit width. The sides of this triangle (Figure 2.11c) are measured not in units of length but in strange units; here the vertical side is in inches (8 ft i.e. 96 inches) of water depth, but the horizontal is in lb/in². This is quite ok in mechanics. So, the area of this triangle is

$$1/2 \bullet [(3.456 \text{ lb/in}^2) \bullet (96 \text{ in})] = 165.8 \text{ lb/in}$$

The wall carries 165.8 lb/in of width. This means that a wall with a width of 100 inches supporting water at a depth of 8 ft carries a horizontal load of 16,580 lb. Notice that we multiply units like any numerical factors so that $(\text{lb/in}^2) \cdot \text{in} = \text{lb/in}$ – i.e., one of the two in the denominator cancels the one in the numerator.

Flow Velocity

The motion of solid bodies is easy to conceptualize and to analyze mathematically. We can track visually the trajectory of a tennis ball (Figure 2.12). And we can calculate mathematically

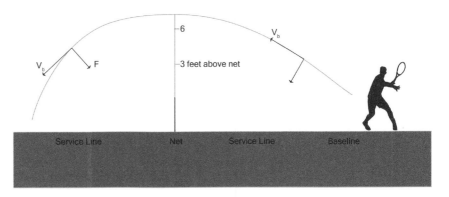

Figure 2.12 Path of a tennis ball, source: author's figure

Figure 2.13 Airplane velocity vectors showing deceleration, source: author's figure

its motion if given its initial velocity (magnitude and direction) – i.e., its velocity the moment it leaves the tennis racket.

Velocity is the physical quantity that indicates how fast an object is moving, expressed as the ratio of the distance traveled over the time it takes to cover this distance. We measure the velocity of a vehicle in miles per hour. But in mechanics, we more often measure velocity in other units of length and time, like feet per second (ft/s), meters per second (m/s), etc. The time rate of change of velocity is called acceleration. When your car accelerates, you can watch its speedometer and record its velocity change as time elapses. Acceleration is a measure of how fast the velocity is changing and is measured in, say m/s per second – i.e., m/s^2. Deceleration is just negative acceleration, as shown schematically in Figure 2.13. This figure shows that in 40 s, a deceleration of 1.5 m/s will reduce the speed of the aircraft from 70 m/s to 10 m/s.

Velocity is a **vector**. It is characterized by both a magnitude and a direction. It is not enough to state that the velocity of a plane is 70 m/s. To have a complete understanding of its velocity, we need to know its direction, north or west, etc. Like all vectors, velocity can be denoted by an arrow, with the length of the arrow denoting its magnitude with some scale and the direction of the arrow indicating the direction of the motion of the object as in Figure 2.12. Notice that in this figure, the velocity vectors are tangent to the trajectory of the ball.

The motion of fluids is more difficult to conceptualize. It is easier to think of fluids as made up of small lumps of fluid that are all packed together. We can then talk about the motion of fluid lumps, and if we can think of these lumps as very small, we can call them **fluid particles**, even though they are not actual solid particles. We can track the trajectory and velocity of fluid particles in the same way we conceptualize and map the motion of

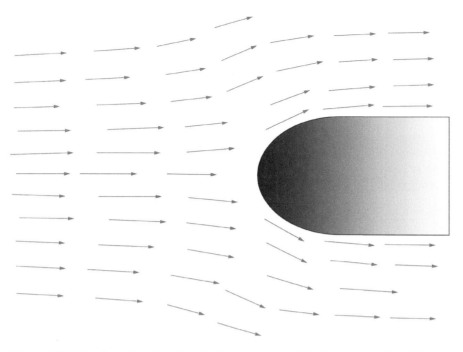

Figure 2.14 Velocity vectors in a flow field over a rounded body, source: author's figure

solids. But it is important to understand that if a fluid occupies a certain space, then at any point in this space there is a lump of fluid – i.e., a fluid particle. And if the fluid is in motion, all its fluid particles are in motion. The flow velocity at a point in space is then defined as the velocity of the fluid particle at this point. Consider, for example, the flow over a solid body, one that is shaped like the nose of an aircraft. At any point in this space, a fluid particle has a certain velocity. And since velocity is a vector, the velocity of each fluid particle can be indicated by an arrow. In Figure 2.14, we present only a few velocity vectors with red arrows. But at any other point in this space, we can define the flow velocity. A space filled with a fluid in motion is called a **flow field**. There are many analytical and experimental methods to determine flow velocities in the space over or around solid bodies. We may refer to some of them in this book.

Streamlines

It is easier to visualize fluid flow in terms of streamlines. A **streamline** is defined as a curve that at all of its points is tangent to a flow velocity vector. Since there is a velocity vector at any point in a flow field, we can define streamlines starting at any point in the field and extend them to the entire domain of our interest. In Figure 2.15, we show the streamlines defined by the velocity vectors of Figure 2.14. If we are provided with only streamlines in a field such as, for example, the patterns of Figure 2.16, we have an accurate picture of the velocity directions but not their magnitude.

The motion of fluids can be visualized or measured experimentally by different methods. One method involves the release of very small and very light solid particles that

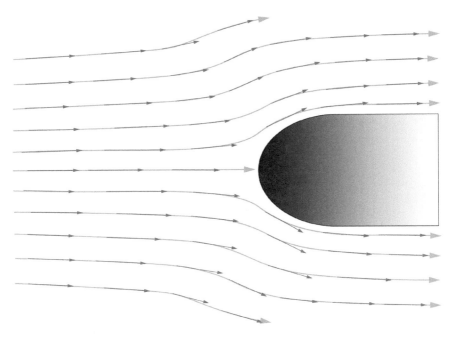

Figure 2.15 Velocity vectors and corresponding streamlines, source: author's figure

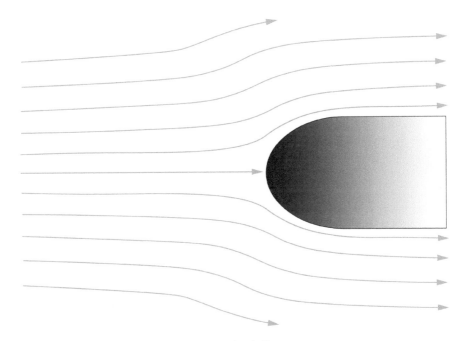

Figure 2.16 Flow streamlines, source: author's figure

are carried by the flow. It can be shown that in most cases, these particles will move along streamlines. The streamlines are therefore trajectories of particles. Snowflakes are good particles that allow us indeed to visualize wind motions, but it is impossible to track and measure their velocity. There are experimental methods that can actually be used to track

Figure 2.17 Streamlines visualized by smoke, source: Wikimedia

special particles that are released in the flow. One of these methods, Particle Image Veloci-metry can capture the instantaneous velocity field if the flow is seeded with actual solid particles. Computer software can then be employed to generate streamlines in terms of velocity vectors thus obtained.

A more practical method, and one that is much less expensive to implement, is smoke or die flow visualization. Smoke is released along a series of sources upstream of the field to be visualized. Typical results are shown in Figure 2.17. This method provides use-ful information. But it is not possible to actually measure the magnitude of the flow velocity with it. Moreover, it is inaccurate if the object about which we visualize the flow is changing in orientation or shape. But it can provide some very useful information, indicating areas in the flow that require special attention as discussed later in this book. For example, we see in Figure 2.17 that the flow (right to left) does not turn down over the car but instead con-tinues downstream. We will define this phenomenon later as **separation**. The flow stops hugging the body and separates from it.

This brief introduction to basic concepts and terms common in fluid mechanics will be used throughout the proceeding chapters.

Chapter 3

Wind and Airflow

Airflow in and around buildings is typically the result of one or two pressure-driving mecha-nisms: wind and thermal buoyancy.

Wind is defined as follows:

> Air in motion; any noticeable natural movement of air parallel to the earth's surface; air artificially put in motion as by an air pump or fan; a strong, fast-moving, or destructive natural current; gale or storm. [1]

Naturally occurring wind is the result of two phenomena: (1) pressure gradients due to dif-ferences in temperature and air density, and (2) the rotation of Earth on its axis

First, when unobstructed or unimpeded, air will move from locations of high pres-sure to low pressure. The ideal gas law relates pressure differences, Δp to changes in tem-perature, ΔT, and volume, V, as in Equation (3.1).

$$\Delta pV = vR\Delta T \qquad (3.1)$$

Here v is the amount of a substance and R is a gas constant. Using Equation (3.1), we can approximate the pressure differences that result from temperature differences in an environ-ment. On Earth at the global, regional, and local scales, temperature differences are driven by the sun and imbalances in the heating of Earth's surface due to differences in albedo (surface absorptance) and incident solar radiation. At the global scale, regions near the equator receive more radiation than at the poles, resulting in the development of macro-cells of air movement that dissipate and move heat outward from the equator. Figure 3.1

DOI: 10.4324/9781003167761-3

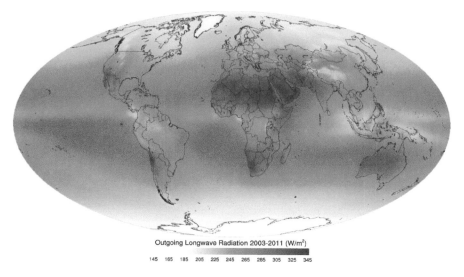

Outgoing Longwave Radiation 2003-2011 (W/m²)

145 165 185 205 225 245 265 285 305 325 345

Figure 3.1 Solar radiation effects and 2D thermal image of Earth, source: Wikipedia

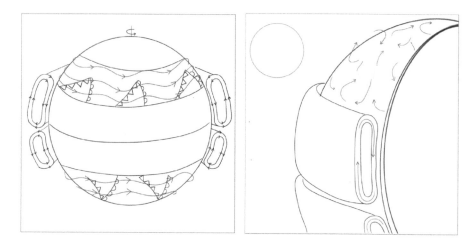

Figure 3.2 Convective airflow cells on Earth, source: author's image

shows the outgoing solar radiation on Earth with a 2D thermal image showing the temperature variation from equator to the poles. These thermal differences create large convective cells of air movement. We experience the air movement due to these cells as the easterlies, westerlies, and trade winds, as shown in Figures 3.2 and 3.3.

At the regional or local scale, wind can result from temperature differences driven by the imbalance in heating Earth's surface such as between the land and water. For example, for coastal locations during the day, the land tends to absorb more solar radiation, warming the air above while the water temperature remains relatively cool, this is shown in the top image of Figure 3.4. The result is temperature and air density differences that produce a surface wind blowing from water inward to the land during the day. At night, the land cools more rapidly than the water due to night radiation, and the scenario is reversed

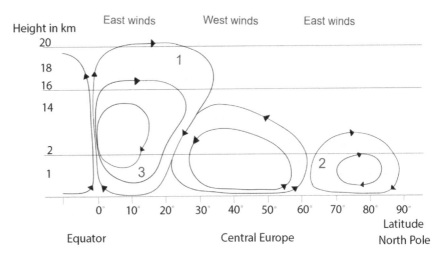

Figure 3.3 Schematic section of air cells on Earth, source: author's image

as the wind blows from land to water, lower image of Figure 3.4. Anyone living or vacationing near the ocean or a large body of water has likely experienced this coastal breeze. At the building scale, this thermal imbalance is used for natural ventilation in courtyard house designs in hot, arid climates.

Historically, before the use of mechanical cooling, hot, arid locations such as Kuwait City (Figure 3.5) relied heavily on these cooling coastal breezes for human comfort. Here, city streets were often laid out to channel these thermally driven breezes throughout the city to provide maximum access to their cooling effect. Wind towers were also implemented and oriented to capture these winds and direct them into the occupied zones of the buildings.

Temperature and pressure differences often result in airflow channeling in locations with varied terrain. In swales and valleys, for example, that do not receive direct solar radiation or during early morning hours, cool, dense air may tend to sink and flow downhill, causing noticeable air movement. When different surfaces receive different amounts of solar radiation, this can also produce localized variations in air density that produces airflow. In the northern hemisphere, for example, south-facing slopes receive more direct solar radiation that can create a thermal plume that can drive local airflow.

Rotational Wind

The second wind-generating force is Earth's rotation. At the equator, due to Earth's rotation, Earth's surface moves at just over 1,000 mph, and this speed decreases proportionally to the cosine of the latitude (the angle from the equator to the latitude as measured from the center of Earth). This rotational velocity sets the atmospheric layer into movement as well. Due to this rotation and drag, near the ground at low latitudes, the easterly trade winds are present while the surface winds at higher latitudes are characterized by the prevailing westerlies. In the upper troposphere at almost all latitudes, the predominant direction of the winds is westerly.

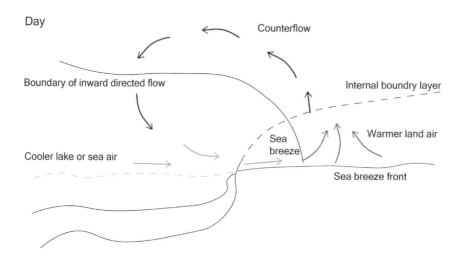

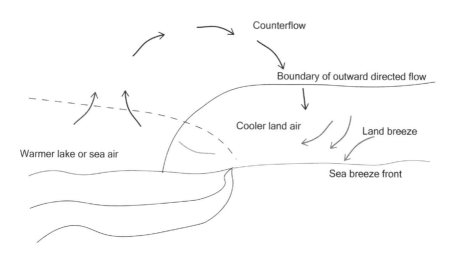

Figure 3.4 Thermal buoyancy-driven coastal breeze, source: author's image

The effect of the rotation of the earth on moving objects is to make them appear as though a force is acting on the objects (Figures 3.6 and 3.7). For example, imagine rolling a ball toward the South from the North Pole. While we would expect the ball to roll due south, it will actually move southward but also to its right, as shown in Figure 3.6. This apparent force that tends to move the ball westward is called the Coriolis force, after the French mathematician G. G. de Coriolis who first treated it mathematically. The effect was used qualitatively a century earlier by George Hadley to explain the trade winds. The reason for introducing the Coriolis force is that an object, such as the wind, moving over the earth's surface with constant velocity relative to an absolute frame of reference appears to be accelerating relative to a person moving with the rotating Earth. The nature of the apparent acceleration and the corresponding force is to turn the wind to the right, as can be seen in

Figure 3.5 Aerial view of Kuwait City, source: Wikimedia

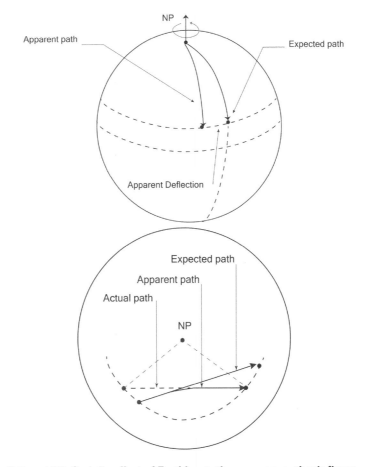

Figures 3.6 and 3.7 Coriolis effect of Earth's rotation, source: author's figure

the previous figures. The result is an apparent force acting at right angles to the motion (of the wind) to produce a deflection to the right in the northern hemisphere and to the left in the southern hemisphere. This is a primary reason for the rotational direction of hurricanes, tornados, and water down a drain. Interestingly, this force is not present at the equator, and here water will flow down a drain without rotation. This rotation often leads to vortices, a flow pattern that will be discussed in the next section.

Vortices

Hurricanes and tornados have rotational airflow. A very simple flow pattern in fluid mechanics is a **vortex**. In its most familiar form, a vortex is characterized by circular streamlines and a velocity that varies with the inverse of the distance from its center (Figure 3.8). This variation is expressed as a function of the radial distance r:

$$V = \Gamma/2\pi r \tag{3.2}$$

Here Γ is a constant, which is essentially a measure of the strength of the vortex. This mathematical expression appears simplistic at first, but it characterizes many eddies in natural flows or in flows generated by man-made equipment. The velocities in a tornado or a hurricane vary indeed very closely to the inverse of the distance from the center. The only difference is that in real life, close to the center, the velocity instead of growing extremely high, decreases and goes to zero. This variation is best demonstrated in a hurricane, in which the wind velocity increases as one gets closer to the core, or equivalently, in case the observer is stationary, the velocity increases as the hurricane is approaching, but very close to the core, the velocity sharply decreases. Inside the core, there is very little motion. This behavior can be seen in Figure 3.9, where we present information on Hurricane Ike, which came ashore with maximum speeds of 110 mph in September 2008. These data were obtained by National Oceanic and Atmospheric Administration during reconnaissance flights (hurricane hunters) and velocity data from the National Weather Service Houston/Galveston WSR-88D radar. In this figure, the wind direction is marked by white arrows, and the wind speed is presented in terms of color contours. The color code

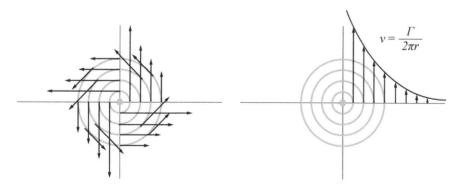

Figure 3.8 An ideal vortex: (a) streamlines and velocity vectors and (b) velocity distribution, source: author's figure

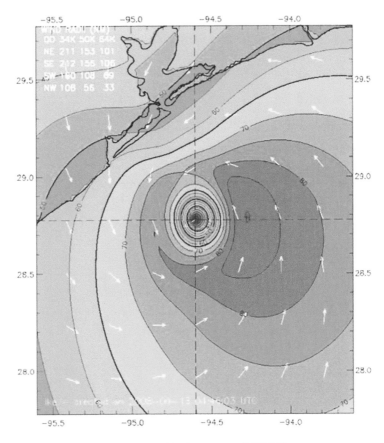

Figure 3.9 Velocity field in Hurricane Ike (courtesy of NOAA)

follows the standard color spectrum, and in this case, high values are designated with red and orange, and low values with yellow, green, and blue. The actual numerical values are marked on the lines defining the domains of color. The core is well defined with yellow, going to green and ending with deep blue, the center of the core.

Vortices are created by large fluctuations of pressure or temperature. But they are also generated when the flow separates over solid surfaces. Typical, common, and well observed are vortices around the corners of a building, so often visualized by the circular motion of dead leaves on the ground.

Regardless of how vortices are generated, they move with the flow. It is easy to think of their motion as if they are floating bodies that drift with the flow or some liquid paint that is carried in a water stream. This is not far from the way vortices behave. But actually, they can affect the flow and also interact with each other and alter their motion. Take, for example, just two vortices shown in Figure 3.10. In Figure 3.10 left, we show two counter-rotating vortices, whereas in Figure 3.10 right, the vortices are co-rotating. Vortex B behaves as if it were a piece of dirt in the flow field of Vortex A. When counter-rotating, Vortex A induces a velocity equal to $\Gamma/2\pi r_0$ up to Vortex B. But now Vortex A is in the field of Vortex B, which induces to it a velocity $\Gamma/2\pi r_0$ also up. As a result, these two vortices propagate with the same velocity upward, along straight lines. In the case of co-rotating vortices, the induced velocities are in opposite direction. The trajectories of these vortices are circular with a diameter of r_0.

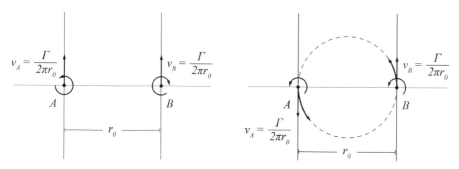

Figure 3.10 Velocities induced by vortices, source: author's figure

Flow Pressure

As discussed in the section on hydrostatics in Chapter 2, fluid elements exert pressure on each other. But this is also true for fluid elements in motion. And as in the case of fluid elements at rest, fluid elements in motion exert pressure on the solid boundaries in contact with them. Just like we did in order to understand the action of the water on a retaining wall, we can imagine separate fluid elements moving together and exerting pressure on each other. We show this schematically in Figure 3.11.

In this figure, we indicate the pressure exerted on Element C from Element B as p_{BC} and the pressure exerted on Element B from Element C as p_{CB}. And based on the principle of action and reaction, p_{BC} is equal to p_{CB}. In the same way, a fluid element moving right next to a retaining wall, say Element D is exerting a pressure p_{DW}, which in this case is the wall, and the wall is pushing back on Element D with a pressure p_{WD}. As discussed earlier, pressure is not a force. Pressure is a scalar, but its action generates forces. These forces are normal to the surface to which the pressure is exerted.

We now see that a fluid moving over a wall, which could be the inside wall of a duct or a pipe, or the outside wall of a structure, exerts pressure on the wall. The structure wall could be the wall of a building, the surface of the roof, the surface of a wind turbine blade, or any other man-made structure. It could also be the surface of an airplane wing or a bird's wing. Just like in the case of a fluid at rest retained by a wall, the pressure exerted by a moving fluid varies with distance along the wall. We indicate such variations in Figure 3.12, where the magnitude of the pressure is shown by the length of black arrows. We emphasize again that pressure is a scalar. This quantity is not associated with any direction. But when pressure is exerted over a surface, it gives rise to forces which locally are in a direction normal to the surface, as shown in Figure 3.12.

Notice that along the front of this solid body, the pressure is shown larger, and it decreases as we follow its contour to the back. We will explain later why the pressure varies in this way over a solid body.

A beautiful example of how harnessing flow pressures can produce desirable forces is the lift generated on wings, artificial or natural. This is shown schematically in Figure 3.13.

As we will explain later, a wing at an angle of incidence generates on its upper side pressures lower than the ambient and on its lower side pressures higher than the

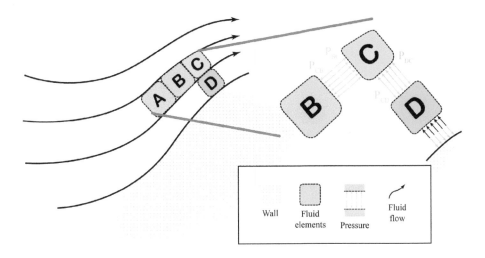

Figure 3.11 Pressures exerted on fluid elements, source: author's figure

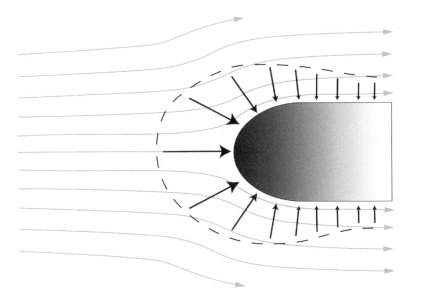

Figure 3.12 Pressure exerted on a solid surface indicated in terms of black arrows, source: author's figure

ambient. The ambient pressure is the pressure of the surrounding medium, such as a gas or liquid far from the object. Engineers and architects employ the terms **pressure side** and **suction side** to denote sides that experience higher and lower than the ambient pressure. A quick observation of the distribution of pressures over the wing of Figure 3.13 indicates that the net effect is a force pointing up. This force is called **lift**. Engineers can calculate such flow forces during the design process, or measure them on laboratory models. Architects could make approximate estimates or employ powerful computer tools to calculate these forces such as lift on a roof during a high wind event. They could also conduct model experiments (Figure 3.14) or collaborate with engineers to run large-scale experiments.

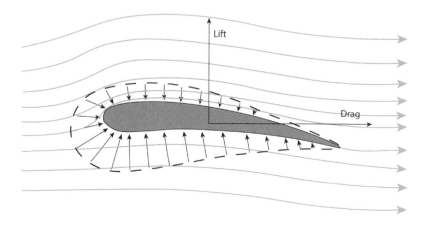

Figure 3.13 Pressures developing over a wing

Figure 3.14 Building wind tunnel model (courtesy RWDI Anemos Limited)

To withstand forces generated by the wind, structures require special reinforcements. This is especially true for geographic areas susceptible to strong atmospheric disturbances, including wind. In extreme cases, like in hurricane or tornado situations, wind forces can cause catastrophic destruction. In other situations, depending on structural poor designs, even moderate winds could damage a structure by lifting its roof or even destroy a large steel structure. An often-cited example of the latter is the Tacoma Narrows Bridge (Figure 3.15). In the chapters that follow, we will discuss the uplifting of roofs and the resonance phenomena that can damage structures exposed to even moderate winds. This discussion will explain how the wind induced larger and larger oscillations of the Tacoma Narrows Bridge structure that resulted in its collapse.

Figure 3.15 Tacoma Narrows Bridge collapse, source: Wikipedia

Continuity Equation

There are many analytical models that describe the motion of fluids, and most of them can be expressed in terms of complex mathematical equations. Some closed-form solutions have been found for only very simple configurations. But modern computer algorithms have been developed to provide practical tools that can be used by designers for a variety of practical configurations. These algorithms are based on models known as CFD (Computational Fluid Dynamics). Today, powerful CFD codes are commercially available.

In this section, we present two of the most basic laws of fluid mechanics, the conservation of mass and Bernoulli's equation that apply to situations such as that shown in Figure 3.13. These can prove somewhat useful in estimating the behavior of flow motions and the pressures they generate over solid bodies such as buildings. More useful to the readers of this book is the insight these equations can provide.

Let us consider the flow through a solid pipe of a general shape that might involve bends and variations of its cross-section, such as a plumbing riser or air duct, as shown in Figure 3.16. The product of the area A at a section, times the velocity there, V is the volume rate of the fluid, often represented by the symbol Q – i.e., $Q = VA$. This is the volume of the fluid passing through this section per unit time. If the velocity is measured in ft/s and the area of the cross-section in square feet, then the multiplication calculated by their product is given in cubic feet per second, ft^3/s. In practice, engineers and technicians measure the flow rate of liquids in gallons per minute. To convert ft^3/s to gallons per minute, one needs to multiply by the factor 448.83.

If we multiply the flow rate by the density of the fluid, $\rho Q = \rho V A$, we get the mass flow rate, and if we multiply by the specific weight, we get the weight flow rate, $\gamma Q = \gamma V A$.

Let the cross-sectional area at stations 1 and 2 in Figure 3.16 be A_1 and A_2, respectively and the velocities and densities at these stations be V_1, V_2 and ρ_1, ρ_2, respectively.

We can now express the **conservation of mass**, one of the fundamental laws of physics, in terms of these quantities. This law states that the mass of a body cannot be destroyed. In our case, this implies that the mass of the fluid that passes through station 1, will be equal to the mass of the fluid passing through station 2.

$$\rho_1 V_1 A_1 = \rho_2 V_2 A_2 \tag{3.3}$$

For all practical purposes, the density of water is constant – i.e., $\rho_1 = \rho_2$. And for air at low speeds, i.e., for speeds less than 50 m/s, we can assume again that $\rho_1 = \rho_2$. We call this flow **incompressible** flow. The factors ρ_1 and ρ_2 then cancel out in Equation (3.3), which then becomes

$$V_1 A_1 = V_2 A_2 \tag{3.4}$$

This law is often referred to as the **continuity equation**. A direct conclusion we can draw from this law is that since the product of the velocity times the area is constant, if one of these quantities decreases along a pipe, then the other must increase. In other words, if the area of the pipe decreases with distance, then the velocity increases.

Bernoulli's Equation

The flow pressure in the pipe in Figure 3.16 varies with the distance in the pipe. In fact, the variations in pressure are connected with the variations in velocity. These are controlled by **Bernoulli's equation.**

$$p_1 + \rho V_1^2/2 = p_2 + \rho V_2^2/2 \tag{3.5}$$

This relationship can be derived mathematically from a basic principle of mechanics, the **conversation of energ**ψ. It is not a relationship obtained by trial and error. And it holds for any incompressible flow, like the flow of water and the flow of air at low speeds. Equations in mechanics must have terms consistent in dimensions. In this equation, each term has units of force over area. So the first term, pressure in English units, must be given in pounds per square inch. If in the term $\rho V^2/2$, each factor is expressed in English units, then its unit works out in pounds per square inch. Thus, this term also has units of pressure. In fluid mechanics, this term is known as **dynamic pressure**. The first term, p, is an as **static pressure.**

Here is now a very important conclusion we can draw for the variation of pressure along a pipe. If as we go from point 1 to point 2 in Figure 3.16, the velocity increases

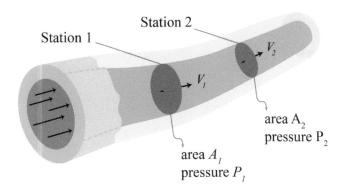

Figure 3.16 Flow through a pipe, source: author's figure

as discussed earlier; then, by virtue of Bernoulli's equation, the pressure decreases, and vice versa. This is called the **Bernoulli effect**. The quantity $p_2 + \rho V_2^2/2$ stays the same, and Equation (3.5) holds. We can draw one more very interesting conclusion by referring back to Equation (3.4). We concluded from this equation that if the area of the pipe decreases, then the pressure decreases. This may sound a little anti-intuitional to some. It indicates that if you step on a garden hose, thus reducing its cross-sectional area, the pressure at this point does not increase, it actually decreases.

If we are dealing with the flow of a liquid in a pipe that involves different elevations, then the equation takes the form

$$p_1 + \rho V_1^2/2 + \gamma z_1 = p_2 + \rho V_2^2/2 + \gamma z_2 \tag{3.6}$$

Here, ρ and γ are the density and specific weight of the fluid, p denotes the pressure at a certain station, V denotes the velocity, and z denotes the elevation. This relationship holds between any two points along the pipe. If we define other points, say a point 3 and a point 4, anywhere along the pipe, we can write the equation as

$$p_1 + \rho V_1^2/2 + \gamma z_1 = p_2 + \rho V_2^2/2 + \gamma z_2 = p_3 + \rho V_3^2/2 + \gamma z_3 = p_4 + \rho V_4^2/2 + \gamma z_4 \tag{3.7}$$

We can now introduce another concept in fluid flow. Consider a flow field much more extended than the flow in a pipe. Consider the streamlines that touch a fixed closed contour. These streamlines form now a tube, which is called a **streamtube**. A streamtube is like a pipe, but its walls are not solid. The flow in a streamtube is constrained by the fluid moving outside of its imaginary wall. Any flow field can be conceptually thought of consisting of imaginary streamtubes (Figure 3.17). The Gulf Stream flowing in the Atlantic Ocean can be thought of consisting of many steamtubes. The velocity and pressure variations along a streamtube obey the fluid mechanics laws of the equations of this section.

The equations presented here cannot predict all the properties of the flow over a body or through an enclosure. More powerful equations known as the Navier-Stokes equations are available. These are complex partial differential equations that defy closed-form solutions – i.e., solutions derived on paper. But these equations can be solved numerically on a computer using CFD.

Figure 3.17 Streamtubes, source: author's figure

Boundary Layers and Separation

Flow **boundary layers** are layers of fluid moving over a solid boundary within which the velocity decreases sharply. Wind moving over Earth forms an **atmospheric boundary layer**, where the wind velocity decreases from its undisturbed velocity level high above the ground to very small values on the ground. In fact, very close to the ground, the wind speed is zero. In the atmospheric boundary layer, the wind velocity is decreased because of the drag induced by obstacles like trees or man-made structures.

On a much smaller scale, but of great interest to engineers and architects, are the boundary layers that grow over structural solid surfaces, like the walls of a building, the body of a vehicle, the wing of an aircraft, or the blades of a wind turbine. To understand this effect, let us consider the flow over a thin, flat plate aligned with an undisturbed free stream, as shown in Figure 3.18. We show in this figure the velocity profile upstream of the plate as uniform. As the stream moves over the plate, its velocity on the plate must go to zero at the surface of the plate. In fluid mechanics, this is referred to as the **no-slip condition**. Indeed, it has been experimentally shown that fluids cannot slip over a solid surface.

The presence of the solid wall induces a slowing down of flow layers moving over it. This effect extends further from the wall as the flow moves over the wall and creates velocity profiles shown schematically in Figure 3.18. The extent of this slowing down is denoted in the figure with the Greek symbol delta, δ. This is the thickness of the boundary layer – i.e., the distance from the wall to the edge of the domain of reduced velocity – shown in the figure with a dashed line. In this figure, we show the free-stream velocity as uniform and equal to U. Far from the wall, the flow velocity takes the undisturbed value of the free-stream velocity.

This example gives us the opportunity to introduce the concepts of **fluid friction** and **shear stresses** in fluids. Let us consider the fluid layer that moves immediately above the plate wall, as shown in Figure 3.19a, highlighted yellow in the figure. It has been shown experimentally that this layer exerts a shearing force F on the plate wall (Figure 3.19b). It is trying to drag the plate with it. By the principle of action and reaction, then the plate exerts an equal and opposite force on the fluid (Figure 3.19c). This means that the fluid is dragging the plate to the right, but the plate is "fighting back" by trying to slow the motion of the fluid. The plate must be secured to a solid ground to hold its position and not be swept with

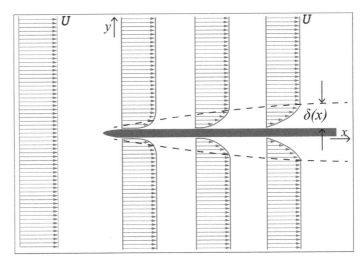

Figure 3.18 A boundary layer growing over a flat solid plate, source: author's figure

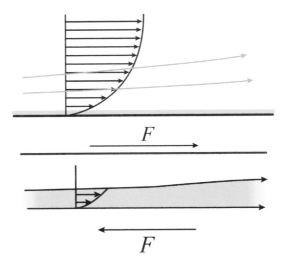

Figure 3.19 Flow layer moving over a solid plate, source: author's figure

the stream. On the other hand, the plate slows the fluid and holds the fluid's speed to zero on its surface.

Forces that are parallel to a surface on which they act are called **shear forces**. And these forces can be distributed just like the normal forces we discussed in earlier chapters. The distributed normal forces we called pressures, and we denoted them by the symbol p. The distributed shear forces are called **shear stresses**, and we will denote them by the Greek letter tau – i.e., τ. Shear stresses are due to friction between layers of flow. Both kinds of distributed forces, normal distributed forces – i.e., pressure – and shear distributed forces – i.e., shear stresses – have dimensions of force per area, and their most common units are pounds per square inch, lb/in^2, or psi.

Let us now consider fluid layers that flow next to each other, layers denoted by different colors in Figures 3.20. The three highlighted layers we call in this example layers A,

B, and C, as indicated in Figure 3.20, are colored blue, red, and yellow, respectively. Let us consider now layer B alone and the shear stresses exerted on it. Layer C exerts on B shear stresses denoted by τ_{CB}, and these stresses are pointing to the left. They are trying to slow layer B. But layer A exerts on layer B stresses τ_{AB} to the right. So, layer A is trying to keep layer B in motion. We see now that it is because of these stresses originating on the plate wall that the flow layers are slowed down and produce the velocity profiles of a boundary layer, as shown in Figures 3.18 and 3.19.

The layers shown in these figures are drawn expanding – i.e., thickening – as they move downstream. This is actually a true feature of boundary layers, and it is due to the fact that if in each layer, which is essentially a streamtube, the velocity decreases as dictated by the continuity equation (Equation (3.4)). The area that the streamtube occupies must increase, and thus it pushes the layer further away from the wall.

The shear forces F of Figure 3.19 and the shear stresses of Figure 3.20 are proportional to the viscosity of the fluid usually denoted by the Greek letter mu, μ, and F is measured in units of lb sec/ft^2. Another quantity often used in aerodynamics is the kinematic viscosity, denoted by the Greek letter nu, ν; it is the ratio of the viscosity of a fluid over its density, $\nu = \mu/\rho$. Kinematic viscosity is measured in units of ft^2/s. The viscosity and kinematic viscosity of water and air at room temperature are given as follows.

Water: μ = 0.0000209 lb sec/ft^2, ν = 0.0000102 ft^2/s
Air: μ = 0.000000376 lb sec/ft^2, ν = 0.000161 ft^2/s

The viscosity of most fluids has been measured and can be found in tables. Fluids of high viscosity like oil or synthetic fluids like adhesives give rise to large values of friction. The viscosity of water is very small and that of air is even smaller. And yet, as will be explained later, they have a strong effect on the forces exerted on different bodies and structures.

The effect of fluid friction – i.e., the slowing down of the flow layers – does not extend far from the wall. This extent is called the thickness of the boundary layer and is often denoted with the Greek letter delta, δ. In most practical cases, the **boundary-layer thickness** is a small fraction of the size of the body over which the boundary layer is growing.

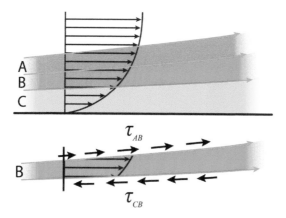

Figure 3.20 Flow layers and shear stresses, source: author's figure

For example, the boundary layer over a wing with a chord of 15 feet has a thickness of a few inches. The boundary layer thickness gets smaller for higher flow speeds.

The thickness of a boundary layer depends on the flow velocity, the length of the body over which the fluid is flowing, and the viscosity of the fluid. These quantities are grouped together in the product called the **Reynolds number**, which is usually denoted by the symbol Re:

$$Re = UL/\nu, \tag{3.8}$$

where L is a typical length of a body and U is the free-stream velocity. This number is a dimensionless number, just like the ratio of two lengths or the ratio of two weights. Indeed, the numerator in Equation (3.8) has units of velocity times length – i.e., in English units, ft/s times feet, which makes it ft^2/s. The kinematic viscosity has also units of ft^2/s, and as a result, the ratio UL/ν is dimensionless.

For laminar flow over a flat plate, the thickness of a boundary layer was found to be approximately

$$\delta/x = 5/\sqrt{Re}. \tag{3.9}$$

Here x is the coordinate measuring the distance along the plate measured from its leading edge, and the Reynolds number depends on the length of the plate, L. For example, for air moving over a plate of length 1 ft, at a speed of 1 ft/s, the Reynolds number becomes

$$Re = UL/\nu = (1\ ft/sec)(1\ ft) / (0.000161\ ft^2/sec) = 6200, \tag{3.10}$$

and the thickness of the boundary layer at the end of the plate is

$$\delta = x\left(5/\sqrt{Re}\right) = (1\ ft) \bullet 5//\sqrt{6200} = 0.0634\ ft. \tag{3.11}$$

Viscosity and kinematic viscosity of air and water are very small, and thus shear stresses are also very small. But for large, long bodies, shear stresses could produce a significant effect. Consider, for example, airflow over the fuselage of an aircraft. In a frame of reference attached to the aircraft, air is moving over the aircraft, and it generates shear stresses τaw exerted from the wind to the aircraft, as shown schematically in Figure 3.21. These stresses add to a considerable net force to the right, which adds to the drag on the fuselage.

The shear layers growing on a building wall or even over the surface of a moderate-size vehicle generate shear stresses that are practically insignificant. But their effect plays a very important role in controlling the forces exerted on a structure when these layers separate from the wall.

This phenomenon can be described nicely with the example in Figure 3.22. In this figure, we show the flow in a duct, which diverges in shape. Diverging ducts are called **diffusers**. Boundary layers grow on the inside wall of the duct. But the flow cannot make the sharp turn along the expanding part of the wall and tends to keep flowing in its original direction. As a result, some parts of the fluid reverse direction and generate pockets of **recirculation**. The boundary layer lifts off from the wall at a point along the wall. This is

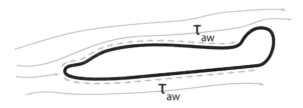

Figure 3.21 Shear stresses τ_{aw} **exerted from the air to the wall of an aircraft fuselage,** source: author's figure

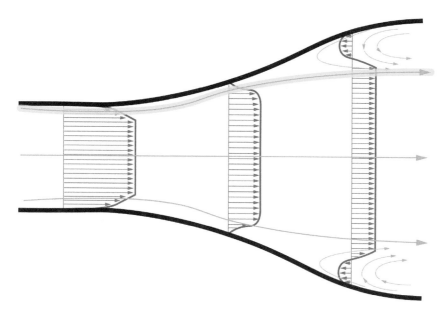

Figure 3.22 Flow in a diverging channel, source: author's figure

called the point of **flow separation**, or just **separation**. The attached boundary layer is highlighted in the figure in yellow. The boundary layer detaches from the wall, as shown with the yellow highlight. Underneath this separated layer forms a region of recirculating flow. Right next to the wall, the flow is in the opposite direction of the oncoming stream. The separated layer is called a **free-shear layer**.

Let's now consider wind over the surface of the earth, which plays the role of the flat plate we discussed previously. As wind moves from location to location, its speed varies not only due to the previously described pressure differences, where dense pressure gradients produce higher wind speeds, but velocity also varies vertically from Earth's surface. In nearly all convective flow situations when the moving fluid (wind) flows over a surface (Earth) a velocity profile develops where the speed reduces to zero at the surface due to drag and varies vertically until it reaches the velocity of the free air stream at some height above the surface. The result is a nonlinear **boundary layer profile** similar to that shown in Figure 3.23. For more open terrains, such as large bodies of water or fields, the thickness of this boundary layer is typically less than about 700 ft (from ground up to the free air stream velocity). For locations with varied terrain, moderate or dense vegetation or obstructions

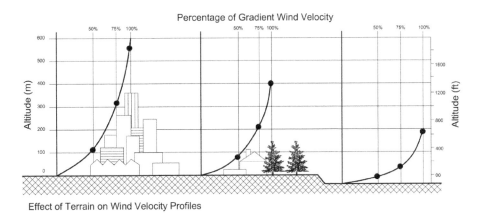

Effect of Terrain on Wind Velocity Profiles

Figure 3.23 Effect of terrain on boundary layer [3], source: author's figure

such as buildings, the layer height might be 900 to 1500 ft. In their book *UDouble Skin Facades*, Oesterle et al. [2] suggest that the mean wind speed at a given height may be approximated by the equation

$$v_n = \left(h_n / 10 \right)^\alpha v_{10} \left[m/s \right] \ . \tag{3.12}$$

Where:
v_n is the wind speed at altitude n in [m/s], h_n is the height of n in [m], and α is the roughness factor. They further suggest that the roughness factor α is the standard value derived from the typical (boundary layer) curve. It will be roughly 0.16 for extensive areas of water or open terrain, roughly 0.20 for forest areas, and roughly 0.22–0.24 for evenly developed urban fabric. This is especially valid for strong wind conditions where a building is not situated in a valley or hollow or set among a group of structures of similar height.

Turbulence

The layers of the free stream of Figure 3.20 move in a well-ordered way and retain their order when they move over the solid plate. They do not undulate or mix with each other. This type of flow is called **laminar** flow. But at some point, over the plate or any other solid boundary, and for no apparent reason, the flow may develop first some unsteady undulations, and then it breaks up into large and small eddies that roll over each other as shown schematically in Figure 3.24. This type of flow is called **turbulent**. Laminar flow can turn into turbulent flow in many other circumstances. Typical is the example of the flow entering a duct or a pipe, as shown in Figure 3.25. Again, with no specific interference or obstruction, the flow may turn from laminar to turbulent. This is the case of turbulence first observed by Reynolds, who released dye at the entrance of the pipe, watched it stay laminar for a while (Figure 3.25 top), and then turn into turbulence (Figure 3.25 bottom).

Figure 3.24 Boundary layer over a solid surface may break up into multiple eddies, source: author's figure

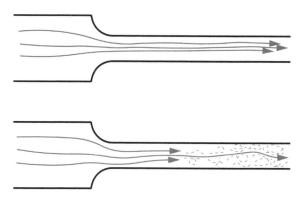

Figure 3.25 Flow in a duct may transition to turbulence, source: author's figure

The turbulence that develops in a boundary layer is confined to the thickness of the boundary layer. We show above the boundary layer of Figure 3.24 some smooth streamlines to indicate that the flow outside the boundary layer is laminar. The character of a boundary layer affects the shear stresses that it exerts on the wall. The stresses of a turbulent boundary layer are much larger than the stresses of a laminar boundary layer. Still, their overall effect on the structure is very small. But the laminar or turbulent character of a boundary layer has a significant effect on the location of separation. Turbulent boundary layers tend to stay attached to the wall for longer distances – i.e., they delay separation. As a result, the wake generated by a turbulent boundary layer is smaller than the wake generated by a laminar boundary layer, as described later in this chapter.

The transition to turbulence in a boundary layer moving over a solid wall or the flow in a duct appears to be spontaneous. Actually, it is caused by very small disturbances, which literally let the flow "trip" over itself. Large differences in the velocity of fluid layers that move parallel to each other in what we called laminar flow make this flow unstable. In other words, the faster moving layers tend to turn toward the lower moving layers and in the process form eddies.

But turbulence can also be generated by large obstructions. The flow behind natural and man-made structures induces the formation of eddies of large size that roll around each other and sustain each other for long distances downstream. This part of the flow is called the wake of the obstruction. Trees and buildings obstruct the flow of the wind and give rise to large eddies. The wakes of these obstructions mix with each other and form what we referred to earlier as the atmospheric boundary layer.

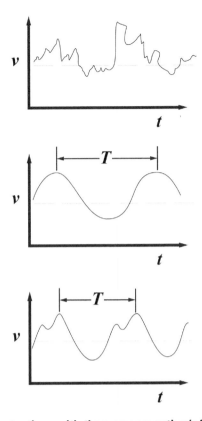

Figure 3.26 Velocity fluctuations with time, source: author's figure

The basic feature of turbulent flow is that it involves large-scale fluctuations of velocity and pressure. This fact can most easily be demonstrated by a velocity measuring device or by a pressure sensor. We will discuss later the tools available for the measurement of the flow velocity at a point in the flow. For the time being, let us call such a tool a **velocimeter**. If we insert a velocimeter in the pipe of Figure 3.25, then at stations close to the entrance to the pipe, the instrument reads a constant value – i.e., a reading that does not change with time. But if inserted further downstream, then the reading fluctuates with time, as shown in Figure 3.26 top. These fluctuations do not show any organization. To explain the turbulent character of the flow, it would help to indicate exactly the opposite – i.e., "organization" in a signal. Fluctuating signals can be characterized by periodic organized fluctuations – i.e., increases and decreases of a fixed duration. This is demonstrated in Figure 3.26 middle. The time between consecutive maxima or minima or any other repeated characteristic of the fluctuation is fixed. This is called the **period** of the fluctuation, and it is usually denoted by the symbol T and is measured in time units, like seconds or minutes. The inverse of the quantity is called the **frequency** of the fluctuation, often denoted by the symbol f, and thus $f = 1/T$. Frequencies are measured in units of inverse time – i.e., 1/s – which is called a **Hertz**.

A signal could contain two frequencies like the one shown In Figure 3.26 bottom and still be considered well organized if the pattern is repeated in time. Another character

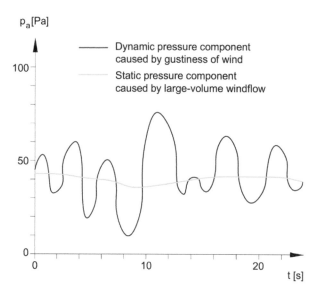

Figure 3.27 Static and fluctuating pressure variations, source: author's figure

of organization is the **amplitude** of the fluctuation. In Figure 3.26 middle and bottom, the maxima always reach the same height. Now we can see that the signal of Figure 3.26 top lacks organization because it displays many frequencies, and the amplitude of the fluctuation varies randomly. This is the most basic characteristic of turbulence. The flow in a turbulent boundary layer, or in the wake of a structure display, fluctuations with large frequency ranges and great amplitude variations.

In part due to this boundary-layer phenomenon and nearby obstructions, typically, wind velocity is not constant and fluctuates above and below the mean (Figure 3.27). As a result, the wind pressure on building surfaces is comprised of two components: static and fluctuating. The static component is the result of large-scale air movements, typically resulting in gradual changes in pressure. Static pressure primarily leads to pressure differences between the windward and leeward sides of the building. The fluctuating pressures, on the other hand, are the result of wind turbulence and cause short-term changes in wind speed and direction caused by eddies in the airstream as it flows past a building. A quantitative expression of these wind unsteady fluctuating components is the **turbulence intensity**.

Turbulence intensity (Tu) is an indicator of the magnitude of the turbulent flow and is defined as the ratio of the standard deviation of the air speed (SD_v) to the mean air speed (v). Turbulence intensity may also be expressed in percent (i.e., Tu = [SD_v/V] 100). Typical turbulent intensity values range from about 10% to 60%.

Separated Flow and Wakes

As described earlier, fluid flow cannot make sharp turns, and thus it separates over any bluff body. The location of separation over a smooth curved surface depends on the characteristics of the boundary layer, which in turn are controlled by the free-stream velocity and the kinematic viscosity of the fluid. A case that has been extensively studied by engineers is

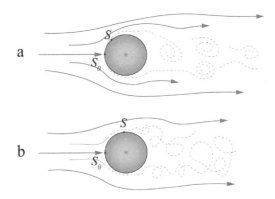

Figure 3.28 Flow over a circular cylinder, source: author's figure

the flow over circular cylinders (Figure 3.28). A point of separation is denoted in this figure with the letter S. At low speeds, the location of separation is closer to the front of the body (Figure 3.8 a). At higher speeds, the point of separation moves downstream on the body (Figure 3.8 b).

What is special about the flow over a circular cylinder is that it makes analytical and experimental investigations very convenient. But the flow features over this body are universal and represent the flows over any structure, and therefore the study of this flow serves the purposes of this book.

Another distinct feature of the flow over bluff bodies is that as the flow approaches the body, it bifurcates so that some of the fluid flows over the body, and some flows under the body, as shown in Figure 3.28. The streamline that separates the two streams, the one that goes over the body and the one that goes under the body, actually ends on the surface of the body. The velocity along this streamline decreases as the fluid approaches the body until right on the surface of the body the velocity goes to zero. The flow "stagnates," and the point where this special streamline ends is called the **stagnation point** and is denoted in the figure with the symbol S_o.

There is something very special about the stagnation point. The pressure on the wall of the body at this point is the largest pressure that the flow exerts on the body. Bernoulli's equation can be used to actually calculate the stagnation pressure. Equation (3.5) is repeated here for convenience:

$$p_1 + \rho V_1^2/2 = p_2 + \rho V_2^2/2$$

Let us choose point 1 far upstream where the flow is undisturbed. We can denote the pressure and the velocity there with p_∞ and V_∞. We choose point 2 to be right at the stagnation point. The velocity there is zero, and we will call the pressure there **stagnation pressure** and denote it as p_o. Bernoulli's equation then becomes

$$p_\infty + \rho\, V_\infty^2 / 2 = p_o . \qquad (3.13)$$

Here p_∞ is the ambient static pressure. The quantity $\rho\, V_\infty^2/2$, which is measured in units of pressure is called **dynamic pressure**.

Working with dimensionless quantities offers the advantage of not depending on the choice of units. Moreover, dimensionless quantities represent better the physical meaning of a quantity. For example, using the height of a building as a unit and presenting dimensionless lengths as a ratio of the dimension of interest to the height of the building gives a better image of the quantity under consideration. The dynamic pressure, $PV_\infty.^2/2$ is a good choice as a pressure unit and is often used to define dimensionless pressure. And if we choose gauge pressure, $p - p_\infty$ instead of absolute pressure, then a dimensionless pressure becomes

$$C_p = (p - p_\infty) / (\rho V_\infty.^2 / 2) \hspace{2cm} (3.14)$$

This is called **pressure coefficient** and is denoted with the symbol C_p. We can now present the pressure over the surface of the circular cylinder in terms of the pressure coefficient (Figure 3.29). In this figure, the C_p is plotted against the angle θ, which is measured away from the stagnation point.

The pressure coefficient at the stagnation point takes the value of 1. Following along the surface of the body away from the stagnation point, the pressure coefficient on the wall is decreasing, in fact becoming negative in the aft part of the body. This means that the pressure is getting even smaller than the ambient pressure p_∞. The coefficient decreases until the point of separation, and from then on, it remains approximately constant at a low value usually referred to as the **base pressure** and denoted by p_B. The location of separation essentially defines the beginning of the low-pressure region. Over the aft part of the body, approximately for values of θ larger than 90°, the graph lines are horizontal and negative, indicating that the pressure there is less than the ambient pressure and constant. This is the base pressure. The level of the base pressure depends on the condition of the boundary layer that flows over the forward part of the body. If the boundary layer is laminar, then the flow separates a little earlier than if the boundary layer is turbulent. As a result, the base pressure for laminar flow is lower than the base pressure for turbulent flow.

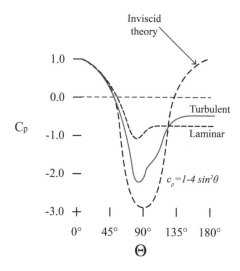

Figure 3.29 Pressure coefficient over a circular cylinder, source: author's figure

When a boundary layer separates, it retains its basic character of having high velocities on one side and zero, or very low, velocities on its other side. Such layers are called **free-shear layers**. The highlighted layer in Figure 3.30 denotes the attached boundary layer, which after separation becomes a free-shear layer. Free-shear layers define a space downstream or above the body where the flow velocity and pressure are very low. This domain of the flow is called the **wake**. We found a similar pattern of a free-shear layer delineating a separated region in the internal flow of Figure 3.22.

The location of separation thus defines the extent of the wake and therefore the area over the body subjected to low base pressure. This is where the Reynolds number emerges again. The location of separation depends on the free-stream velocity, the size of the body, and the fluid kinematic viscosity combined in the form of a Reynolds number.

For flows over bluff bodies, the Reynolds number is based on a typical dimension of the body. In the case of a circular cylinder, the diameter D is used. So, the Reynolds number is defined as $Re = V_\infty D/\nu$. After a large number of experiments, it was found that for Reynolds, numbers below about 500,000, the flow separates at about $\theta = 85°$, where θ is the angle measured from the stagnation point. For Reynolds numbers over 500,000, the flow separates at about $110°$. In Figure 3.31 left, we present a sketch of the flow with separation at $\theta = 85°$, and in Figure 3.31 right, we provide a sketch for separation at $\theta = 110°$. Note that the separated flow region for the first case is much wider than the second case.

In the frames of Figure 3.32, we present schematically the pressure distribution that corresponds to the flows of Figure 3.31 left and right. The pressure distribution on the

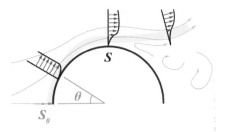

Figure 3.30 When a boundary layer separates from the wall, it becomes a free-shear layer, source: author's figure

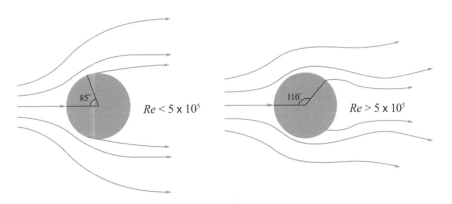

Figure 3.31 Flow separation over a circular cylinder, source: author's figure

front part of the body is more or less the same for both high and low Reynolds numbers. But in the aft part of the body, the pressure for the higher Reynolds numbers is much lower than the pressures for low Reynolds numbers. Since the pressures on the windward side are larger than those on the leeward side, one can see that the net effect must be a result-ant force to the right. This force is called **drag**.

In cases of flow over solid surfaces with corners, the flow separates at the cor-ners. In Figure 3.33 top, we show the flow over a rectangular structure such as a building and separation at the windward corner S. But again, the flow over a specific structure may create different separation patterns. In Figure 3.33 top, the flow is shown separating over the windward corner, but the separation line meets the flat roof at a point we will call the **point of reattachment**, indicated by the symbol R. This creates a pocket of recirculating flow. Further downstream, the flow separates again and generates a much wider wake.

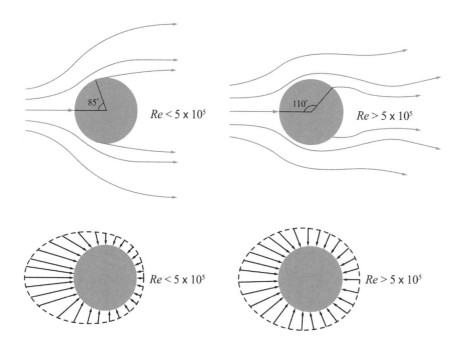

Figure 3.32 Pressures over a circular cylinder, source: author's figure

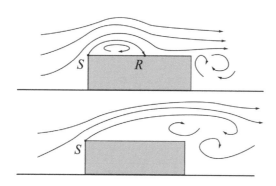

Figure 3.33 Flow over a cornered structure, source: author's figure

Here the Reynolds number again emerges. For higher Reynolds numbers, the separation line that leaves the body at the windward corner may not reattach to the body and thus creates a wider wake, as shown in Figure 3.33 bottom.

When an undisturbed fluid (wind) encounters one or more slender objects, the flow streams are deflected (Figure 3.34). If the angle of the obstructing surface is slight or near parallel with the direction of the prevailing wind and the object's depth shallow, the flow stream may be only slightly deflected and the flow left generally unaffected, as in Figure 3.34 top. However, as the obstruction becomes larger and at a greater angle to the stream, the flow condition is affected as air is diverted over and around the object. The result is a zone downstream of the obstruction where flow is agitated, chaotic, and difficult to predict, similar to the flows over bluff bodies described in Figure 3.34 bottom. This is characteristic of a turbulent flow condition. As the fluid continues beyond the obstruction at some distance downstream, the disturbed air is re-entrained with the prevailing wind, and the flow returns to its original more laminar condition. The area of the turbulent zone downstream of the obstruction depends on the nature and angle of the obstruction and the initial velocity of the wind.

When wind encounters a rectilinear-shaped building, three general airflow patterns result, also discussed in Figure 3.33. First, on the windward face (into the wind), the flow stagnates and splits into an upwardly diverted stream (upwash) and a downwardly diverted stream (downwash) that creates a vortex, shown in red in Figure 3.35. The flow around the stagnation point is relatively slow. If the oncoming stream does not include turbulence, then in this region pressure is steady.

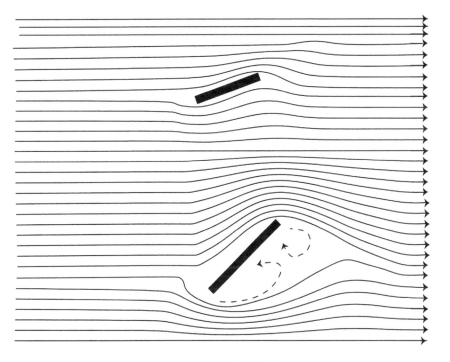

Figure 3.34 Obstructed flow field, source: author's figure

Second, behind the building on the downwind or leeward side develops the wake, the zone of recirculating flow, also often called the wind shadow. This is typically characterized by turbulent flow with large velocity fluctuations and relatively low flow velocities when compared to the prevailing wind. This zone is also typically at lower pressure than on the windward side, similar to Figures 3.33 and 3.35.

Finally, as the wind flows over the roof of the building, the air streams are compressed, and wind speed accelerates, as in Figures 3.33 and 3.35. Bernoulli's equation (Equation (3.5)) suggests that as air speed increases (near the windward corner of the roof), pressure decreases. Because of the increased wind speed over the top of the building, whether the separated flow re-attaches or not, a low-pressure zone develops on the roof.

In Figure 3.36, we display schematically the external pressure distributions that correspond to the flow patterns of Figure 3.35. We should emphasize again here that static pressure is always positive – i.e., compressive. Fluids cannot sustain tensile forces. But the net pressure exerted on a wall will be the difference between the external pressure and the internal pressure. So when the internal pressure is higher than the external pressure, the net effect is suction, which is denoted in Figure 3.36 with negative values. Another point of

Figure 3.35 Flow over a building, source: author's figure

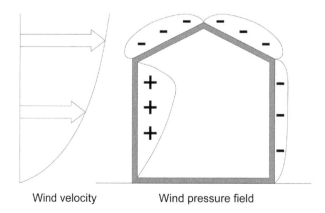

Wind velocity   Wind pressure field

Figure 3.36 Section view of wind pressures on a building, source: author's figure

Figure 3.37 Roof damage by very strong wind, source: Wikipedia

interest in this figure is the presence of the atmospheric boundary layer. Very low near the ground, the velocity is small and goes to zero on the ground. There the net pressure, the difference between external and internal pressure, goes to zero, or it could even become negative.

In high wind conditions, suction on the roof is a serious concern, as this low pressure can damage the roof (Figure 3.37) or even lift the roof off of the structure of the building. This is a common failure condition during hurricanes and is discussed in detail in Chapter 5.

Pressure distributions over structures of different shapes and dimensions can be calculated by computational methods or by wind tunnel experiments. These are becoming more and more available to practitioners. But they are still quite expensive. Empirical methods to estimate wind loads on structures are available.

The ASCE 7–10 for wind-induced pressure acting on the walls of an enclosed building by Equations (3.15) and (3.16). [4] Here,

$$P_w = qGC_p - q_i(GC_{pi})$$ (3.15)

Where:

G = gust effect factor

C_p = external pressure coefficient

(GC_{pi}) = internal pressure coefficient

q = velocity pressure, in psf, given by the formula

$$q=0.00256K_zK_{zt}K_dV^2 \qquad (3.16)$$

Here:

$q = q_h$ for leeward walls, side walls, and roofs, evaluated at roof mean height, h

$q = q_z$ for windward walls, evaluated at height, z

$q_i = q_h$ for negative internal pressure,

$(-GC_{pi})$ evaluation, and q_z for positive internal pressure evaluation $(+GC_{pi})$ of partially enclosed buildings but can be taken as q_h for conservative value.

K_z = velocity pressure coefficient

K_{zt} = topographic factor

K_d = wind directionality factor

V = basic wind speed in mph

The important relationship from Equations (3.15) and (3.16) is that, similar to the negative uplifting pressure on the roof due to the Bernoulli effect, the positive windward pressures are proportional to the square of the velocity, thus doubling the wind speed results in a fourfold increase in pressure on a windward facing exterior surface. Understanding these relationships can allow for better utilization of strategies such as natural ventilation or design conditions to reduce the unwanted wind-driven pressure distribution on the building.

As described earlier, when wind encounters a surface perpendicular to its predominant direction, a zone of pressure higher than the ambient develops that pushes on this surface, and the tendency is for air to enter the building through cracks and openings. If the inward flow of air is uncontrolled and unwanted, we term this **infiltration**. *Infiltration is the flow of outdoor air into a building through cracks and other unintentional openings and through the normal use of exterior doors for entrance and egress.* Infiltration is also known as air leakage into a building. The Jacob House, shown previously, represents a design strategy to minimize infiltration by reducing the exposed surface area in the direction of the prevailing wind through earth berming.

Typically, on the downwind or leeward side, where a negative pressure zone develops (wind shadow) air would be drawn out of the building through cracks and openings. When the wind is nearly perpendicular to the windward wall, the side walls and roof also experience negative pressure and air exfiltration, in part due to the acceleration of the wind around the building resulting in a Bernoulli effect.

To design for effects of airflow around buildings, wind speed and direction frequency data are necessary. The ASHRAE suggests,

> The simplest forms of wind data are tables or charts of climatic normals, which give hourly average wind speeds, prevailing wind directions, and peak gust wind speeds for each month (Figure 3.38). This information can be found in sources such as the Weather Almanac (Blair 1992) [5] and the Climatic Atlas of the United States (DOC 1968) [6]. A current source, which contains information on wind speed and direction frequencies, is the International Station Meteorological

Climate Summary CD from the National Climatic Data Center (NCDC) [7] in Ashville NC.

Computer-based tools such as Climate Consultant, available through the School of Arts and Architecture at the University of California at Los Angeles [8] are useful for processing these data into an easy-to-understand and visualize format. Wind speed data are typically displayed as bar charts, histograms, or frequency distribution as a wind rose or wind wheel (Figure 3.39).

It should be noted that wind can be a highly variable phenomenon. Using a single prevailing wind direction for design can cause serious errors, so care should be taken to understand the specific on-site conditions. For any set of wind conditions, one direction always has a somewhat higher frequency of occurrence. Thus, it is often called the prevailing wind, even though winds from other directions may be almost as frequent. As previously introduced, wind speed can be highly fluctuant and velocity may vary significantly above and below a mean value, particularly in turbulent flow conditions.

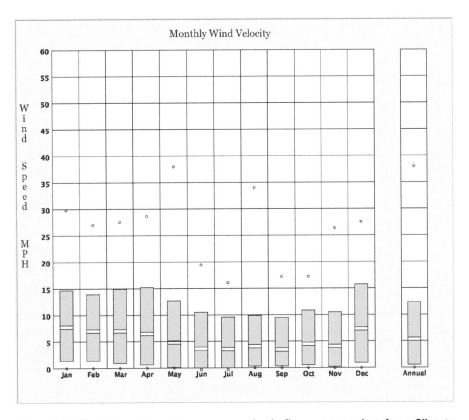

Figure 3.38 Monthly wind data, source: author's figure, screenshot from Climate Consultant

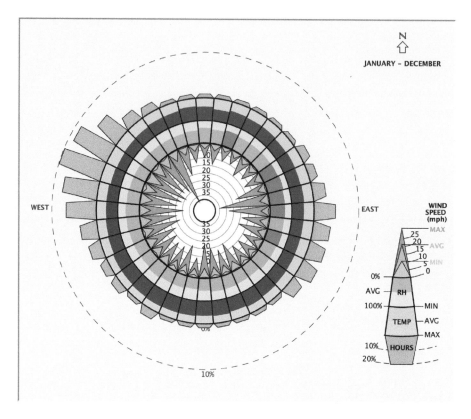

Figure 3.39 Wind rose (wheel) diagram, source: author's figure, screenshot from Climate Consultant

Flow in Three Dimensions

Vortices develop over any structure exposed to the wind. The development of such vortices strongly affects the pressure distributions over the structure surface, as well as ventilation characteristics and acoustic effects. Engineers and architects should antic-ipate such developments and seek to analyze and predict the corresponding pressure distributions. This can be done by employing computational methods or running wind tunnel experiments. Such methods can be implemented by engineering specialists. But the architect should anticipate the special need to assign properly the design research. The architect should also be able to communicate technically with the specialists and understand and evaluate technical results. Along this end, we will introduce in this section another two concepts of fluid mechanics – namely, **skin-friction lines** and **stream traces**.

So far, we described in some detail the flow over bodies in two dimensions, which we will call **two-dimensional flow**. The world of course is not two-dimensional. What we discussed so far is flow over cylindrical bodies that extend very far in the direction normal to the direction of the flow. Let us first define carefully the concept of such bodies. A

cylindrical body is defined by straight generators parallel to each other that are touching a closed contour. A circular cylinder is a good example since its surface can be defined by the generators that touch a circle, and are parallel to the axis perpendicular to the plane of the circle (Figure 3.40 left). Another cylindrical body could be a section of a body with a rectangular cross-section, as shown in Figure 3.40 right. A cone is not a cylindrical body because its generators are not parallel to each other

Back to two-dimensional flow now. If a free stream is approaching a cylindrical body with its free-stream velocity perpendicular to the generators of the body, then the flow that develops is the same along planes normal to the axis of the body. This is what we call two-dimensional flow. All the flows discussed and depicted schematically so far in this book are two-dimensional flows. If we want to expose the character of such flow in three dimensions, then the flow over the cylinder of Figure 3.28 can be represented in Figure 3.41, showing the free-shear layers also called **vortex sheets** as developable surfaces.

Developable is a surface that can be developed by reshaping a flat sheet. One can think of the vortex sheets of this figure as made up of flat sheets of paper that were rolled in these shapes. Vortex sheets released over solid bodies actually roll in the form of individual vortices, which play an important role in the development of pressures over a structure.

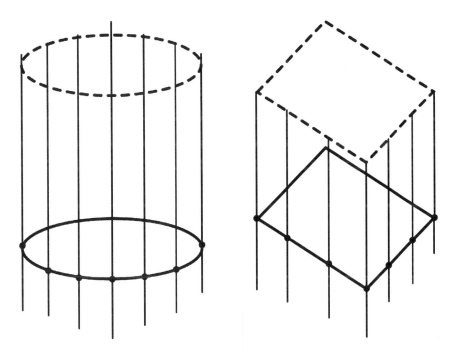

Figure 3.40 Cylindrical bodies defined by straight generators, source: author's figure

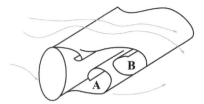

Figure 3.41 Two-dimensional flow over a circular cylinder approached by a stream normal to its axis, source: author's figure

Figure 3.42 An aircraft equipped with a delta wing, source: Wikipedia

Let us consider now three-dimensional flow, which is flow developing over arbitrary three-dimensional bodies. A very interesting flow of interest to aerodynamics is the flow over a delta wing. Some aircraft have wings shaped in the form of a delta (Figure 3.42). We will show that the character of the flow over a delta wing is very much related to the flow over rectangular structures, which are very often elements of architectural design. In its most simple form, a delta wing is an isosceles triangle cut out of a flat plate.

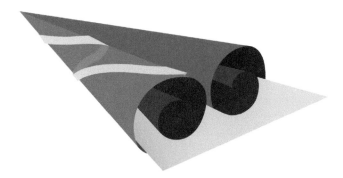

Figure 3.43 Vortex sheets developing over a delta wing, source: author's figure

Consider now a plain delta wing at an angle of attack. The flow over such a wing separates right along its leading edges. Vortex sheets similar to those released along the edges of a rectangular structure roll over the wing, as shown schematically in Figure 3.43. These structures, often called **delta-wing vortices**, generate very large speeds over the wing, which in turn induce very low pressures on the upper surface of the wing. The net resultant of these low pressures is the force that lifts the aircraft, which is essentially "hanging" on these vortices. We will see that delta wing vortices form and shed over many other structures like buildings and bridges

Consider now the flow over a rectangular structure. For simplicity, we will start with a simple cube inserted in a flow with its free-stream velocity perpendicular to one of the faces of the cube.

The flow over the cube separates as expected along the three edges of the front face. Vortical structures develop over the roof and the side walls. Depending on the Reynolds number, these structures could break down to turbulence and extend further downstream. Moreover, again depending on the Reynolds number, the flow may reattach on these three side surfaces. The back end will be completely immersed in the separated flow. But again, the flow may reattach on the ground close or far from the structure. Another interesting flow development is flow separation on the ground upstream of the structure. This gives rise to a vortical structure that wraps around the building and hugs the low part of its walls. This is called a **horseshoe vortex**, as shown in Figure 3.44.

So far, we presented hand-drawn sketches drawn by architects and engineers with experience of wind flows over different structures. We can now present the results of accurate numerical calculations. In Figure 3.45, we present numerical results for a rectangular structure exposed to the wind. These were obtained by using a Large-Eddy Simulation Code El Okda [9]. Streamlines are presented along three planes, the midplane, a plane a quarter width from the side plane, and a plane along the side plane. It should be noted that these lines are defined based on the velocity components along these planes. But there are velocity components out of their planes. In this Figure, we can see separation on the

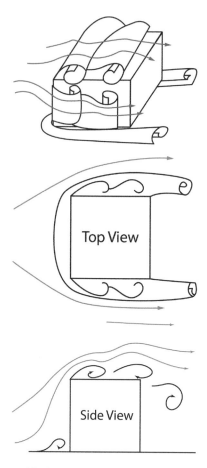

Figure 3.44 Flow over a cubical structure with a stream approaching normal to its front face, source: author's figure

ground upstream of the structure, which gives rise to the horseshoe vortex. We also see the extent of the separated region over the roof and the separated region on the aft side of the structure. By observing the extent of the recirculating patterns in different planes, one can see the extent of separated regions in the spanwise direction.

Pressure distributions over this body were also calculated using the same code, and are presented in Figure 3.46. A color code on the left of this figure is used to display numerical values of the pressure coefficient. This is the pressure coefficient calculated based on the gauge pressure. The largest possible value for pressure is the stagnation pressure, the pressure around the stagnation point. In terms of pressure coefficients, this value is equal to 1. Negative values indicate pressures lower than the ambient pressure. The front surface is completely immersed in red and yellow, indicating pressures larger than the ambient. Pure red regions indicate a point of stagnation, in this case a line of stagnation, or

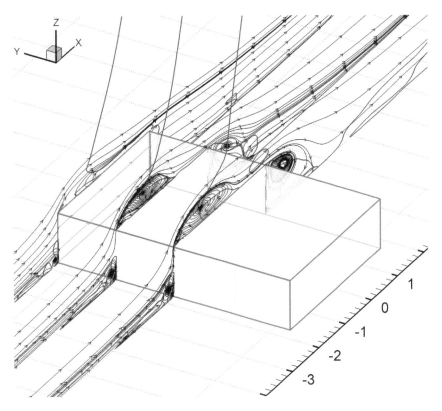

Figure 3.45 Streamlines along a rectangular structure immersed in a stream normal to its front face, source: El Okda [9]

points very close to it. Color coding on side surfaces and the roof indicate negative pressure coefficients as expected for areas of separated flow.

The qualitative information displayed in Figure 3.46 may guide an architect in modifying a design to meet certain requirements. But if needed, these numerical results can be used to provide quantitative information. One can request the calculation of the forces exerted on a face of a structure. Moreover, a designer can see the effect of an opening on the structure. For example, an open window on the front surface and an open window on the side will lead to a forceful wind draft through the internal space, entering on the front and escaping on the side.

There is another kind of information provided in this figure. The lines marked on the structure surface are the **skin-friction lines**, which are defined as the limits of streamlines infinitely close to the walls. These calculations also provide the flow direction infinitely close to the wall and indicated along the skin-friction lines. These are the streaks that fresh paint on the wall will display if the body is immersed in a stream. In fact, this is a technique employed by experimentalists to expose skin-friction lines. A special paint is applied to a model, which is then inserted in a wind tunnel.

Numerical calculations of skin-friction lines actually provide more specific information about the flow. The directions marked on the skin-friction lines can help us identify lines of stagnation, lines of reattachment, and regions of back flow. For example, the skin-friction lines on the roof indicate a region where the flow moves backward, in the opposite direction of the oncoming free stream, which is consistent with the streamline patterns of Figure 3.45 and shown in Figure 3.46.

A different set of lines can help the user understand the nature of complicated flow patterns. These are the **stream traces or streak lines**, which are essentially the trajectories of particles released in the flow as shown in Figure 3.47. For steady flow, these coincide with streamlines, but in the general case of flows that involve dynamic phenomena, like the shedding of vortices, stream traces are distinct and provide special information not otherwise available.

The user of the code can choose the location in the field where particles can be released. In Figure 3.48, we display the streak lines originating along a plane a little to the

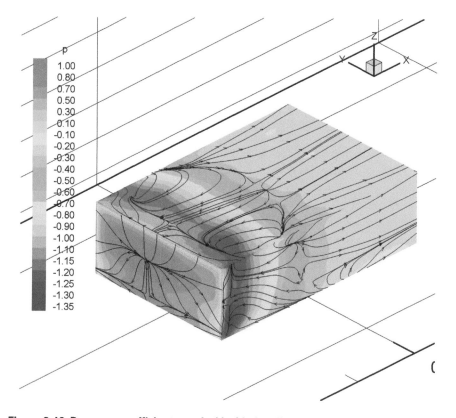

Figure 3.46 Pressure coefficients and skin-friction line on a rectangular structure immersed in a stream normal to its front face, source: author's figure

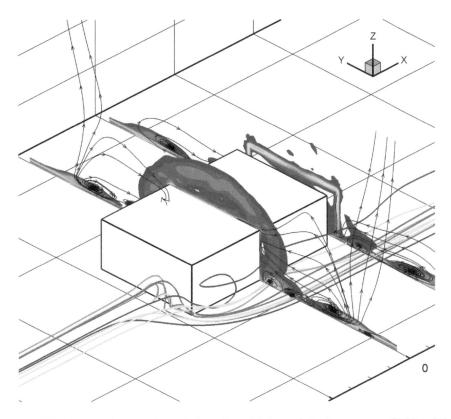

Figure 3.47 Stream traces released along the midplane of the flow, source: El Okda [9]

right of the midplane of the flow. The green and the red lines originate very close to the plane of symmetry, i.e., close to the middle of the front face, and are therefore directed down as they approach the body, and eventually are captured in the horseshoe vortex, and are directed around the body. A blue line reaches the body close to its side wall and is captured by a sidewall vortex. But the green and yellow lines move around the body not entangled in any vortices. In Figure 3.48, we observe stream traces originating on a plane along the extension of the plane of a side wall. These are entrained in sidewall vortices or roof vortices.

Let us consider now the situation whereby the free stream is approaching the structure along a bisector of the edges of the roof as in Figure 3.49. Now the front side faces are inclined by an angle of $45°$ with respect to the oncoming flow. The flow remains attached to these two faces. But it separates along the two edges of the roof. In fact, the roof presents to the stream a configuration very similar to a delta wing, and indeed the vortices released along these edges are identical to delta wing vortices. The low pressure generated below these vortices is powerful. They are often the sources of serious damage to the roofing material.

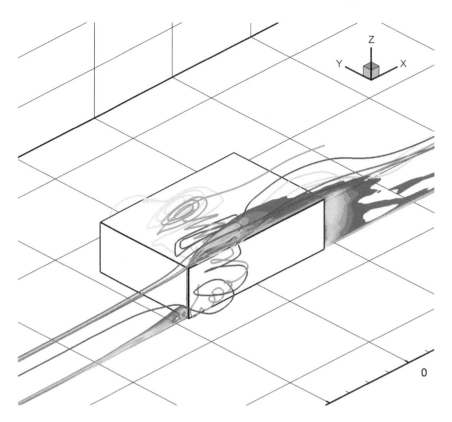

Figure 3.48 Stream traces released along a plane parallel to a side wall, source: author's figure

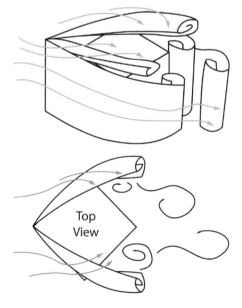

Figure 3.49 Flow over a cube with a stream approaching along a bisector of its top face, source: author's figure

Vortex Sheets and Vortex Shedding

In a very rough approximation, we could represent boundary layers growing over a solid surface in the form of a train of vortices. This is because across a boundary layer we have a large velocity difference. So, the boundary layer growing over a body, like the circular cylinder of Figure 3.30 can be represented by a chain of ideal vortices, as shown in Figure 3.50. At separation, the boundary layers become free-shear layers, and they can again be simulated by a train of ideal free vortices, as shown in this figure. Since real vortices drift with the flow, the present simulation is appropriate. Vortices drift along the solid wall, they are released at separation and then interact with the flow and with themselves. The vortices on the upper part of the flow over the cylinder are spinning in the clockwise sense, and the vortices in the lower part are spinning in the counterclockwise sense. Based on the mechanism described in the previous section and denoted in Figure 3.10, the vortices released on the upper part tend to make the train of vortices roll around each other. This effect is better illustrated in the lower part of Figure 3.50. There the counterclockwise vortices roll to form a cluster, which actually resembles a very large vortex consisting of the individual vortices that rolled around each other. We call this vortex A in Figure 3.50. The growth of vortex A tends to block the vortices on the other side to continue feeding in the cluster on the other side. As a result, vortex B (Figure 3.50) loses its feed from the solid body and is released as a large vortex in the wake of the cylinder. In this way, vortex clusters fill up the wake with vortices and extend far downstream from the body.

The vortex clusters become essentially large-scale vortices, which retain their strength and assume a well-ordered spacing. This was well captured by the numerical simulation of the wake of a circular cylinder shown in Figure 3.51. In this figure, clockwise and counterclockwise vortices propagate downstream. This simulation was obtained for low Reynolds numbers. But these patterns persist for Reynolds numbers as high as about 500,000. These wake patterns are known as the **von Karman vortex street**.

Vortices are formed and shed over any bluff body, including rectangular bodies, which are of greater interest here. The phenomenon is known as **vortex shedding**.

Vortex shedding is present over bodies of all sizes, even the largest on Earth, like islands. In Figure 3.52, the wake of the Juan Fernandez Islands off the coast of Chile is shown. Apparently, vortical structures extend high in the sky and penetrate the clouds over the ocean. It is the disturbance of these clouds that shows the presence of vortex shedding in this figure.

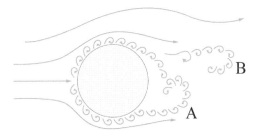

Figure 3.50 Chains of ideal vortices simulate attached boundary layers and free-shear layers, source: author's figure

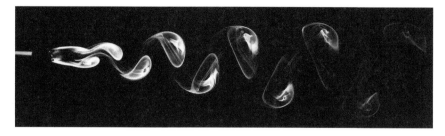

Figure 3.51 Numerical simulation of large-scale vortices forming in the wake of a cube, source: Wikimedia

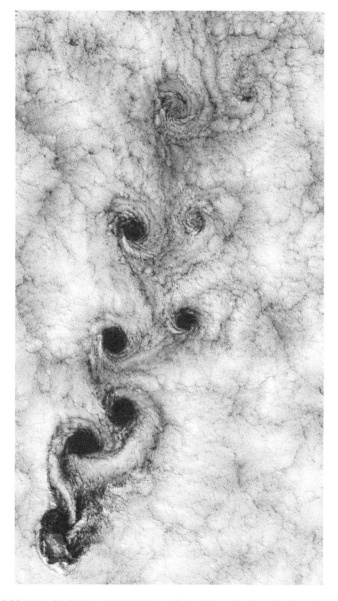

Figure 3.52 Vortex shedding downstream of an island in the Pacific Ocean, Source: Wikimedia

Vortex shedding effects influence greatly the development of loads on structures. The alternate vortex shedding over a solid body leads to asymmetries in the flow. And flow asymmetries imply asymmetric pressure distributions. A significant effect of these asymmetries is that they fluctuate with the frequency of vortex shedding. This means that the pressure on the walls of the structure oscillates in time, which induces fluctuating loads but also creates acoustic effects.

Another significant characteristic of the phenomenon of vortex shedding is that, for a certain configuration, its frequency is fixed and depends only on the free-stream velocity, U. A very large number of experiments has indicated that for a circular cylinder of diameter D, this frequency is given by Equation (3.17).

$$f_s = 0.2U/D = U/5D \qquad (3.17)$$

This phenomenon is consistent and repeatable, and the accuracy of this formula is accepted by scientists so that instruments have been developed using this property in order to measure the velocity of a stream. To this end, a small cylinder is inserted in the flow, instrumented to measure the frequency f_s of the fluctuating loads exerted by vortex shedding. The stream velocity then is $U = 5 f_s D$.

References

[1] Merriam-Webster (2022) *Wind Definition*. Available from: www.merriam-webster.com/dictionary/wind [Accessed October 2019].

[2] Oesterle, E., R.D. Lieb, M. Lutz and W. Heusler (2001) *Double-Skin Facades*. Munich Germany, Prestel Verlag, p. 112.

[3] Melaragno, M.G. (1982) *Wind in Architectural and Environmental Design*. New York, NY, Van Nostrand Reinhold.

[4] ASCE 7–10 (2010) *Minimum Design Loads for Buildings and Other Structures, ASCE/SEI 7–10 for Wind Induced*. Reston, VA, American Society of Civil Engineers.

[5] Bair, F.E. (1992) *The Weather Almanac*, 6th edition. Detroit, MI, Gale Research Inc.

[6] DOC (1968) *Climatic Atlas of the United States*. Washington, DC, U.S. Department of Commerce.

[7] NOAA (2022) *Meteorological Climate Summary CD from the National Climatic Data*. Available from: www.wpc.ncep.noaa.gov/international/scs/MECCA.html [Accessed November 2019].

[8] SBSE (2022) *Climate Consultant*. Energy-Design-Tools: Climate Consultant. Society of Building Science Educators sourced from UCLA College of Architecture and Urban Design. Available from: http:www.energy-design-tools.aud.ucla.edu/.

[9] El Okda, Y.M., 2005 EXPERIMENTAL AND NUMERICAL INVESTIGATIONS OF THE EFFECTS OF INCIDENT TURBULENCE ON THE FLOW OVER A SURFACE-MOUNTED PRISM, PhD dissertation, VA Tech

Chapter 4

Design for Wind Resistance

There are two fundamental responses to aeroform; one being a formal expression of the resistance to wind-driven forces, and the other an expression of the utilization of the wind, such as through natural ventilation. Wind blowing over buildings generates effects that play an important role in architectural design. Moderate winds generate pressure distributions on walls and roofs that could damage the structure. They also affect the flow of air through windows, and thus the efficacy of natural ventilation. Atmospheric turbulence induces dynamic loads that can increase the damaging effects of wind on structures. Unsteady loads are also generated by the interaction of the wind with a structure and give rise to strong fluctuating pressure distributions. These are induced mostly by the phenomenon of vortex shedding. It is due to this phenomenon that large steel structures like the Tacoma Narrows Bridge could collapse even in moderate winds. Strong winds generate similar pressure distributions, but their effect could be catastrophic. The damages created by tornados and hurricanes are very familiar, but little effort has been directed toward designing structures that would mitigate their effect. In this chapter, we discuss wind flow patterns over structures and the corresponding pressure distributions.

An elegant and award-winning architectural example of aeroform for wind resistance is the 30 St Mary Axe building, also known as the Gherkin, located in London. Design by Norman Foster and Partners, and opened in 2004, this 41-story structure incorporates a triangulated perimeter structure referred to as the "diagrid" (Figure 4.1), along with active mass damping, to increase the rigidity of the structure and thus reduce wind-driven sway. [1]

DOI: 10.4324/9781003167761-4

Figure 4.1 30 St Mary Axe by Norman Foster and Partners, source: Wikimedia

Architects need the tools that will provide the pressure distributions over a building. These tools are based on fundamental principles of fluid mechanics described in the previous chapters. Some of these tools are simple and easy to employ. Their accuracy is moderate, but they are very useful in the first steps of an architectural design. More advanced tools like wind tunnel testing or CFD are available, but they are time-consuming and expensive. At advanced levels of the design, collaboration with engineers may prove helpful, and here is where understanding of fluid mechanics principles and the potential of engineering predictions will require direct communication between architects and engineers.

In Chapters 2 and 3, the fundamentals of wind-driven pressures on buildings were introduced. Recall that on the windward side of a building, wind creates positive pressure that is proportional to the square of the incident wind velocity, thus as the velocity doubles, the pressure increases by a factor of four, $2^2 = \mathbf{4}$. A very simple tool that could give an approximate value of the pressure on a wall facing the wind is given by Equations (3.15) and (3.16). These equations give a good rough estimate of the pressures that develop over a wall exposed to the wind. But these pressures are not uniform. They vary over the wall, much like the pressures over a circular cylinder discussed in Chapter 3. These variations depend on the way the flow develops over the structure.

In cases of flow over solid structures with corners, the flow separates at the corners. In Figure 4.2, we show the flow over a rectangular structure. The flow over a specific structure may create different separation patterns. In Figure 4.2 top, the flow is shown separating over the windward corner S, but the separation line meets the flat roof at a point called the **point of reattachment** and is indicated by the symbol R. This creates a pocket of recirculating flow. Further downstream the flow separates again and generates a much wider wake. Here the Reynolds number again emerges. For high Reynolds numbers, that is higher velocities or larger dimensions, or both, the flow over the windward corner may not reattach to the body and thus create a wider wake, as shown in Figure 4.2 bottom.

Here we should emphasize again that pressure is always compressive, it is always exerted toward the solid surface on which it is applied. Recall the definition of gauge pressure, p as the difference between a pressure at a point p_s and the atmospheric pressure, $p = p_s - p_a$. It is the gauge pressure that engineers and architects use, and most of the pressure measuring devices produce. So now, we can talk about negative pressure, which would be pressure less than the ambient atmospheric pressure and indeed they exert suction on a wall.

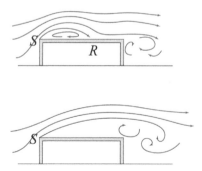

Figure 4.2 Flow over a cornered structure, source: author's figure

On the windward side of a building, pressure conditions result in an inwardly directed force that pushes on the building. On the leeward side, side walls, and roof the pressure fields are typically negative, exerting suction pulling outwardly on building elements, such as walls and windows. Recognizing and responding to these forces through design presents opportunities for the expression of aeroform while avoiding detrimental wind effects. For example, shortly after the building opened, the Hancock Tower in Boston experienced problems when panes of glass were dislodged from the building due to oscillations in thermal stress on the glass panels and the negative rather than positive pressures on the glass that "pulled" the glass from its frame.

The flow over a structure with a pitched roof could generate forces on the roof either up, away from the structure, or down. This depends on the slope of the roof. Consider the pressure generated over a structure with a pitched roof, shown schematically in Figure 4.3 left. In this figure, we show that absolute pressures are exerted on the outside of the structure, as well as the internal absolute pressures. In Figure 4.3 right, we show gauge pressure, which is calculated by subtracting the inner pressures from the outer pressures. Now the wind effect becomes clear. We see that on the roof and on the leeward side of the structure the effect is suction. This is the effect that in extreme situations can actually cause the catastrophe of lifting the roof, sucking out windows or doors, or creating serious structural damage.

The wind pressures on a building will, in part, depend on the characteristics of the surrounding environment. For example, the height and distance to nearby landforms, vegetation, or buildings will influence the atmospheric boundary layer conditions and height of the wind profile while potentially sheltering or funneling wind to the building site. Specific wind effects should be studied at the site or in a wind tunnel while general effects due to the surroundings are shown by the following Equation (4.1).

$$V/V_{met}=Kz^a \qquad (4.1)$$

Where: V = wind velocity at elevation z (m/s)

$V_{met.}$ = Meteorological wind speed at a height of 10 m

z   = elevation measured from the ground

Constants K and a depend on the surrounding terrain, as shown in CIBSE given in Table 4.1. [2]

Figure 4.3 Flow over a pitched-roof structure, source: author's figure

Table 4.1 Terrain wind factors

Terrain	K	a
Open flat country	0.68	0.17
Country with scattered wind breaks	0.52	0.20
Urban	0.35	0.25
City	0.21	0.33

Site Response – Reducing Positive Pressurization

The first opportunity to acknowledge the aforementioned wind conditions through architectural form is through understanding and adaptation to the local site conditions. Through either on-site observation or modeling studies in a wind tunnel or CFD computer simulations, the local wind flow patterns can be determined and used as informants to design solutions. For example, the nestling of the building into the terrain of a sloping site can act as protective arms diverting the unwanted prevailing winter winds over and around the building, thus reducing the pressures on the walls.

As introduced in Chapter 1, built in 1944, Frank Lloyd Wright's Jacobs House in Middleton, Wisconsin, is a constructed example of his hemicycle concept where a semicircular plan building opens to the southern sun for warmth and to the north is bermed into the terrain to divert the harsh winter winds up and over the building (Figure 4.4). A wind rose for Madison, Wisconsin, is shown in Figure 4.5, indicating the higher frequency of winds from the northwest. Nestling the building in the earth to the north as a strategy for wind protection against these winds suggests architectural opportunities for telluric expression where a building arises from the earth, a common gesture by Wright, such as at Falling Water, to let the building emerge from the site.

Similarly, airflow over and around vegetation or nearby landforms and nearby buildings might be used during the site planning phase. Due to their dense leaf structure, conifer or evergreen trees create a wind shadow downwind of the prevailing wind. When grouped together, these trees have the potential to produce a significantly protected zone where wind speeds may be reduced to only one-fourth of their original velocity, as in Figure 4.6. Depending on the wind speed and height, this protective zone may extend from two times the height of the trees to as much as 15 times or more this height. Siting the building in this protective zone relative to the prevailing winter winds can significantly reduce the wind-driven pressures on the building enclosure and reduce infiltration, and subsequently lower heating energy consumption in cold climates. Wind speed reductions due to obstructions are shown in Figures 4.6, 4.7, and 4.8.

Similarly, nearby buildings can act to funnel or divert winds to/around a proposed building. Like a dense group of trees, a building presents an obstruction to the airflow. These obstructions can be useful, as when diverting winter winds, or detrimental, as when channeling these same winds or reducing the opportunities for natural ventilation. The addition to the World Trade Center in Amsterdam by Kohn, Pederson, and Fox is an example of

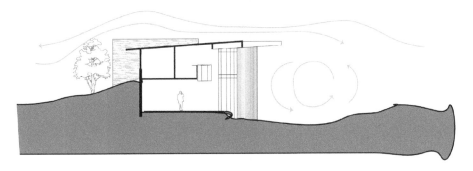

Figure 4.4 Sectional view of Jacobs House, source: author's figure

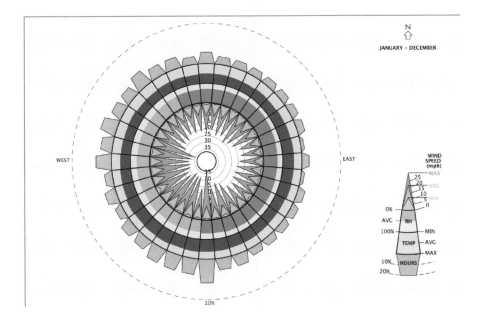

Figure 4.5 Wind rose for Madison, Wisconsin, source: author's figure, Climate Consultant chart output

the design for natural ventilation that took into consideration the relative location of existing buildings and prevailing wind directions. Wind tunnel analysis at Virginia Tech (Figure 4.9) suggested that the existing buildings serve to channel and accelerate the prevailing winds from the north or south over the ventilation openings on the roof of the building addition to induce a draft out of the new atrium. [6]

The images that follow show the plan of the building where the outlined areas are added and the dark elements are existing structures (note north is up on Figure 4.9). The wind rose for Amsterdam (Figure 4.10) showing that prevailing winds often come from the north. The test results suggested that the new atrium will most effectively ventilate when wind is out of the north and northeast directions (Figure 4.9).

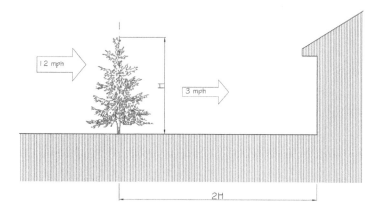

Wind Reduction

(A 20 foot Austrian Pine will reduce 12 m.p.h. winds to 3 m.p.h. for 40 feet on its leeward side.)

Figure 4.6 Wind speed reduction due to vegetation [3], source: author's figure

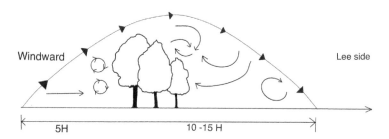

Figure 4.7 Relation of wind shadow length in terms of obstruction height, H, source: author's figure

On the leeward side of an obstructing building, a wind shadow develops with typical low velocity, turbulent flow. Locating new buildings in this wind shadow can reduce the wind-driven pressures on the enclosure, thus reducing effects such as infiltration.

Care should be taken, however, to not locate the building between two existing buildings or near the opening of a large group of trees or hedge row. In this configuration, wind will typically funnel and accelerate at the openings, resulting in greater wind-driven pressures and more infiltration. For openings such as these, wind speeds can be 10% to 20% greater than the ambient velocity.

Also, if natural ventilation is a design strategy, the arrangement of multiple buildings should be carefully considered relative to the prevailing cooling breeze direction. In his book *Environmental Control Systems*, Fuller Moore [7] suggests that a linear arrangement of buildings creates wind shadows (Figure 4.11 left) that minimize the ventilation potential in downwind buildings, while a staggered arrangement (Figure 4.11

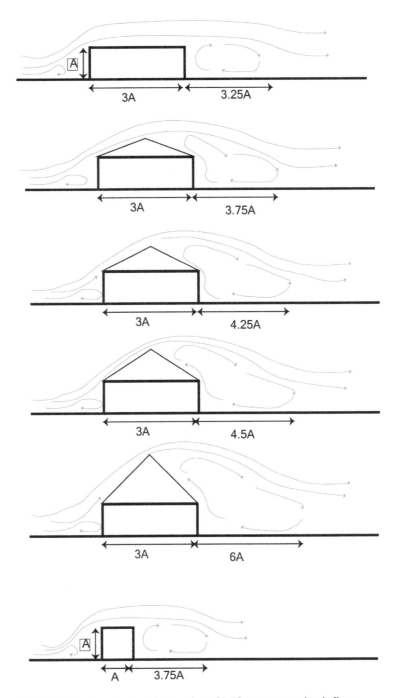

Figure 4.8 Wind shadow due to obstructions [4–5], source: author's figure

right) reduces the wind shadows and increases the opportunities for natural ventilation. In addition, the staggered arrangement presents opportunities to explore the spatial potential of a grid-shifted site plan where buildings might be positioned based on a shifted response to wind and sun while infrastructure such as streets might conform to a preexisting grid.

Figure 4.9 Results of wind tunnel analysis of Amsterdam World Trade Center, source: author's figure

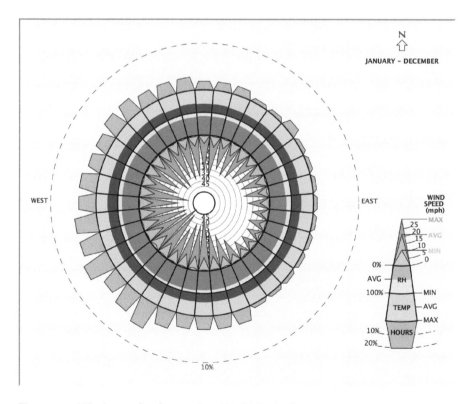

Figure 4.10 Wind rose for Amsterdam World Trade Center, source: author's figure, Climate Consultant chart output

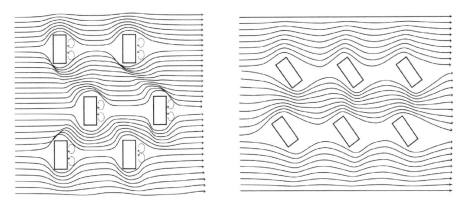

Figure 4.11 Plan view of wind shadow from upwind units and benefit of angled arrangement [8], source: author's figure

Shape Response – Reducing Positive Pressurization

The wind-driven pressures on walls depend not only on the surrounding terrain and velocity, as calculated in Equation (3.16), but also on the angle of incidence between the prevailing wind direction and the outward normal (perpendicular) to the surface. While the exact pressure field will vary from location to location on the wall surface and will depend on other factors such as the geometry of the surface and building height, generally the surface will experience positive (inward directed) surface pressures up to any incident angle less than about 60° from normal. At around this angle of incidence, the wind tends to 'slip' over the surface and accelerate, resulting in zero or balanced pressure conditions. At wind directions of about 70° or higher from perpendicular, the surface is usually under full negative pressure relative to the ambient pressure. These pressure variations are shown in Figure 4.12 similar to Figure 3.29.

A second opportunity for works of architecture to resist or minimize the "push" from wind is through the overall shape of the building. Flat, rectilinear building surfaces when oriented perpendicular to the prevailing wind experience maximum positive pressurization and consequently must offer significant structural resistance to wind-driven forces. Rectilinear surfaces oriented perpendicular to the prevailing winter winds also experience maximum infiltration. Alternative plan geometries, such as triangular shapes, take advantage of angular relationships between the building surface and prevailing wind and reduce the positive pressures on the enclosure.

For triangular shapes, when a vertex (corner) is oriented to the prevailing wind direction, the relationship between the wind direction and surface is no longer perpendicular, and the wind-driven pressures are reduced. As previously shown in Figures 4.12 and 3.29, the average wall pressure coefficient (C_p) is greatest when the wind direction is perpendicular to the wall, decreases to near zero at around a 60° angle of incidence, and becomes negative for wind angles greater than about 70°. By orienting a triangular plan building with the corner into the wind C_p is reduced according to Equations (3.15)

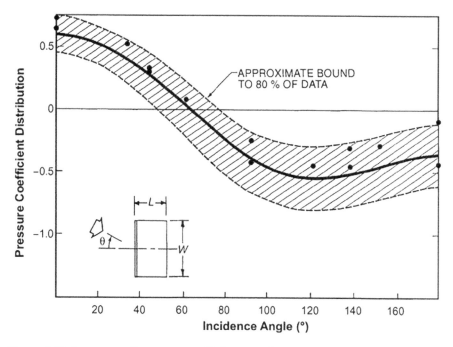

Figure 4.12 Average wall pressure coefficient as a function of wind angle [9], source: ASHRAE

and (3.16). Buildings such as the Commerzbank building in Frankfort Germany by Norman Foster and Partners are designed with a triangular plan that presents this aeroform opportunity (Figure 4.13). It should be noted that even orienting square or rectangular plan buildings such that the walls are not perpendicular to the prevailing winds can reduce the pressure coefficient and resulting forces on the building envelope. The triangular shape, when oriented properly, has the added benefit of reducing the windward downwash condition (see Figure 3.35) and leeward wake-effect turbulent flow that is common with high-rise buildings.

Another application of this pressure-reducing strategy is a circular plan building. The building with a circular plan offers no flat surfaces and therefore minimal perpendicular orientation to the wind. The cylindrical shape of the RWE Building by Overdick and Partners (Figure 4.14) in Essen Germany was derived after wind tunnel modeling that found this shape would reduce the positive wind-driven forces on the building skin. This was important for controlling the ventilation airflow through the double-skin facade.

The KPF Pinnacle Project

Another example of the application of aeroform to a high-rise building is the Pinnacle Project, designed by Kohn Pedersen Fox Associates (London). Also known as the Bishopsgate Tower, proposed to be located on Bishopsgate and Crosby Square in the heart of London,

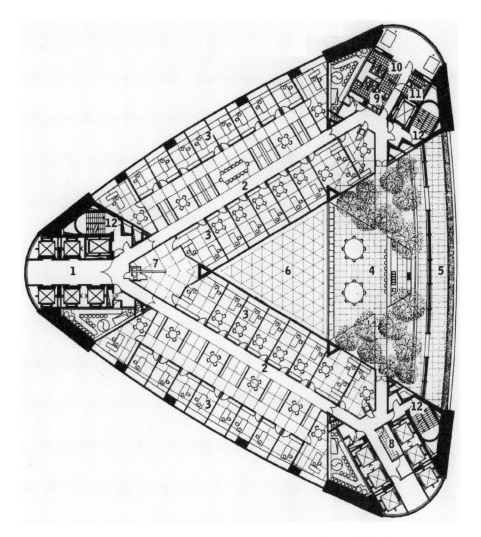

Figure 4.13 Plan view of Commerzbank building minimizing wind-driven pressures (courtesy Uwe Nienstedt Architect)

this building was intended to strengthen the overall character and identity of the emerging cluster of tall buildings in this location and become one of the most significant new buildings in the city. Although it was not constructed, this building was intended to make a substantial contribution to the public realm, opening up the ground level area to pedestrians and linking a number of important urban spaces along Bishopsgate and St Mary Axe (Figure 4.15). The wind management and street-level public domain became significant considerations as the design developed.

As indicated by KPF principal Karen Cook,

> The Pinnacle design, like all tower designs, is informed by wind in its form, structure, details, etc. The design of the form of the building was conceived primarily in response to townscape concerns, as well as client brief goals

such as minimum size tenancy areas and the need to divide the floor in two tenancy areas. Its form comprises inclined flat planes connected by sections of cones, which encourage the wind to flow around the building in a smooth path rather than down to the pedestrian level in a turbulent path. This improves both the pedestrian comfort at street level, as well as reduces the amount of force against the side of the building, so the structure does not have to work as hard.

Figure 4.14 RWE Building Essen Germany by Overdick and Partners

Figure 4.15 Proposed Bishopgate Tower, London by KPF (courtesy of KPF)

She continues to describe the aeroform of the project:

> The site is narrow and the resultant slenderness ratio means that the structural solution is unable to rely on the core for stability against lateral (wind) forces. The perimeter frame thus has to do most of the stability work. The structural engineer introduced a diagonal brace system, which was further refined by parametric modeling and analysis, to remove braces not working very hard and leaving those working hardest. The computer enables the structural engineer to run potentially thousands of series of analyses while by hand the time involved would be prohibitive. In addition to saving construction time and money by erecting fewer elements, the resulting informal appearance reinforces the overall form.
>
> Wind also plays a factor in the cladding treatment and sustainable success of the mechanical system. The skin of the building behaves in a bioclimatic manner, providing sun shading outside the thermal skin for optimum reduction of heat gain (and optimum reduction of carbon footprint). On a tall building the sun shading must be protected from destruction by wind, hence an outer layer of glass protects the sun shading. This outer layer is not for thermal protection, and is open at each edge to allow air to flow through at each office partition module and at each floor. These overlapping outer glass panels we call the "snakeskin." This chamber or cavity also creates a mini buffer zone, which also takes heat away from the facade. Further, the inner thermal glass layer is

opened, to let in fresh air directly. Extensive modeling was done to verify that the wind direction and speed worked with the building form to create adequate pressure differentials on all sides of the building, so that the air would be drawn at appropriate speeds across the interior, neither too draughty nor too stagnant. The dimension of the gaps between the panels was fine tuned to control its precise speed entering the building, and direction of airflow into the space.

Finally, the building design was adjusted numerous times to mitigate wind to improve pedestrian comfort. The snakeskin is pulled away from the body of the tower to create a bottom edge to its tall form and integrate it among the shorter neighboring buildings along the street. This "canopy" is lifted to mark the pedestrian passage through the building, and pulled down to deflect wind into the street. Snakeskin panels are rotated sideways to diffuse wind flow and direct wind down into the path of street level wind, acting as a brake against itself. [10]

The Pinnacle Project was evaluated both computationally and through wind tunnel studies, and the results were used to inform the building shape in response to the aforementioned issues.

Piloti – Reducing Positive Pressurization

Wind forces on buildings can be significant, particularly for buildings in high wind or hurricane-prone locations or high-rise buildings. As will be shown later, high winds not only induce lateral forces on the building; they can also create uplift. A design strategy for reducing these lateral forces is to lift the base of the building off of the ground, thus providing pressure relief from these lateral forces (Figure 4.16). This strategy can often be found in stilt houses in hurricane or cyclone-prone areas of the Caribbean Islands or tropical Pacific regions (Figure 4.17).

For medium height and high-rise buildings, wind-driven pressures can be a significant concern for the structure and facade. As shown in Chapter 2, wind speeds

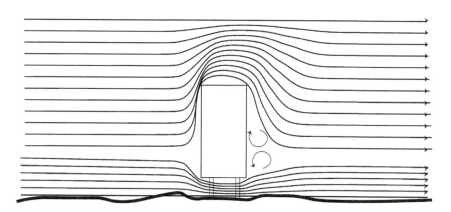

Figure 4.16 Schematic of building raised on piloti, source: author's figure

Figure 4.17 Example of house elevated in tropical climates, source: Wikimedia

typically increase with height above the earth's surface and, consequently, so does the pressure on a building. Wind forces are a significant issue when designing the structural system for tall buildings. This was, in part, the concern for the Pinnacle Tower Project. As with the tropical structures, these pressures can be reduced by lifting the base of the building above the ground. Raising the building through the use of architectural elements such as piloti can relieve some of these pressure forces and potentially reduce the sizing of structural elements. Care should be taken with this design strategy, however, as the wind may be funneled under the building and its speed accelerated, creating a wind tunnel effect on-site.

The piloti as an architectural element is probably best seen in the work of LeCorbusier. LeCorbusier proposed that the piloti was one of the five primary elements in architecture (Figure 4.18). His application of the piloti can best be seen in the Villa Savoye where a 64-foot square box is raised from the ground by 12 round concrete columns. Although not necessarily intended as an aeroform response, for LeCorbusier, the piloti supports a fundamental declaration about architecture and its relation, or lack of, to the site. In his book *Master Builders*, Peter Blake asserts that

> he (LeCorbusier) declared that he was a classicist about form and a classicist about nature. The Villa Savoye is divorced from the ground and raised up against the sky in a precise, geometric silhouette – raised up as if by some giant hand. . ..
> The precise, geometric silhouette of the Villa Savoye permitted no confusion of architecture with nature. [11]

Figure 4.18 Villa Savoye by LeCorbusier

For aeroform, the intention is reversed where indeed the lifting of the building is in response to nature and the force of the wind.

In the Swiss Pavilion in Paris (1930–1932), LeCorbusier suggests yet another architectural interpretation of the piloti. In his book *LeCorbusier and the Tragic View of Architecture*, Charles Jencks suggests that *"here the pilotis support the slab-block as "legs" appropriate to their visual and anthropomorphic function."* He quotes LeCorbusier as remarking, *"The columns of a building should be like the strong curvaceous thighs of a woman."*[12] The works of LeCorbusier are referenced only to suggest that the role of the piloti in architecture can be varied with it potentially being aeroform, an expression of wind pressure relief through design.

Wind and Windows

When designing buildings in response to wind forces, among the most significant facade elements are the windows. As previously mentioned, one of the most notable wind-related window problems was the detachment of some windows in the Hancock building in Boston shortly after its construction. Chapter 16 in "Structural Design" of the 2006 International Building Code [13] requires designers to calculate the design wind pressures for the building, as well as for components and cladding. The main reference for wind design is the American Society of Civil Engineers (ASCE 7), *Minimum Design Loads for Buildings or Other Structures*. [14]

ASCE 7 suggests that wind pressures are influenced by building size, height, geometry, wind exposure, and speed. Calculations for wind design pressure include a variety of other factors such as the intended occupancy and importance of the structure. The ASCE 7 includes three methods for determining wind design pressures for buildings. Probably the most common among these is Method 1 – Simplified Procedures, which uses the following formula for calculating design wind pressure (P_{net}) for enclosure components and cladding:

$$P_{net} = \lambda \times K_{zt} \times I \times P_{net30} \qquad (4.2)$$

Where:

P_{net} = Design wind pressure

λ = Adjustment factor for building height and exposure

K_{zt} = Topographic factor evaluated at mean roof height

I = Importance factor based on building occupancy

P_{net30} = Net design wind pressure for exposure B, at h = 30 ft and for I = 1.0

While the application of this equation and suggested values of the factors are described in full detail in ASCE7 and will not be repeated here, a few key points concerning the factors in the equation should be mentioned. First, the adjustment factor (λ) is based on the mean roof height of the building and the exposure of the building to wind. Generally, the taller the building and the more exposed the site, as previously shown, the higher will be the adjustment factor.

Also, as previously shown, wind speed and the resulting pressures depend on the local topography. When calculating the net pressure on building windows, Equation (4.2) includes a topographic factor (K_{zt}). Evaluated at the mean height of the roof, the topographic factor is equal to 1.00 if all three of the following site conditions are not applicable.

- The hill, ridge, or escarpment that the building sits on is isolated.
- The hill, ridge, or escarpment that the building sits on rises above the height of the surrounding terrain by a factor of two or more.
- The structure is located in the upper one-half of a hill or ridge or near the crest of an escarpment.

Building exposure (B, C, or D) is based on the characteristics of the surroundings. Exposure B is the least severe exposure and includes urban and suburban areas, wooded areas, and other terrain with numerous closely spaced obstructions having the size of single-family dwellings or larger. Exposure C includes open terrain with scattered obstructions having heights generally less than 30 ft. Exposure D is the most severe exposure and includes flat, unobstructed areas and water surfaces outside hurricane regions.

Not all buildings are required to be designed to withstand the highest wind loads. Buildings are rated as to the importance of their occupancy. The ASCE Table I-1: Occupancy Category of Buildings and Other Structures, for Flood, Wind, Snow, Earthquake, and Ice Loads lists occupancy categories that are used to determine the importance factor. Occupancy factors range from IV, the most important, to I, the least important. Essential facilities such as hospitals and emergency facilities, and power plants have importance factors of IV. The importance factor will also depend on the number of occupants with high importance factors associated with greater occupancy. For most buildings, the importance factor will be either 1.00 or 1.15.

Chapter 6 of the ASCE standard provides a series of tables and figures as a basis for determining the net design pressure (P_{net30}). This variable depends on the design wind speed at the project location, the effective area of the component (window), and the building zone in which the component is located. The design wind speed is selected from the wind speed maps included in ASCE Figure 6–1, as shown in Figure 4.19, and is based on a 3-second wind gust, recorded in miles per hour, at 33 ft above the ground in Exposure C. For windows in punched openings, the effective wind area is the area of the window opening in square feet.

The design pressures also depends on the location of the window within the wall. For this, the ASCE standard designates locational zones where zone 4 is for windows located in the center of the facade and zone 5 is for windows located at the corners of a building. Generally, windows located near the building corners (zone 5) experience the highest design wind pressures and may require a different specification from windows located at the center of the building (zone 4).

In the May 2008 edition of *Architecture Record* [15] the Greensboro Public Library (Greensboro North Carolina) designed by J. Hyatt Hammond and Associates is an example of windows designed to balance the desire for abundant daylighting while resisting lateral wind forces from hurricanes that impact the southeast United States. Their solution was a 2-foot square window module that reduces the effective wind area and at the same time "speaks" to the character of the revitalization of historic Greensboro, as shown in Figures 4.20 and 4.21.

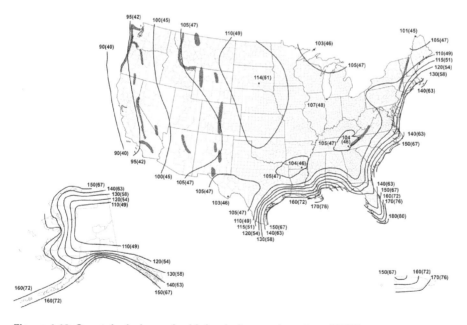

Figure 4.19 Coastal wind map for high wind zones (courtesy ASCE)

Figures 4.20 and 4.21 Greensboro Public Library by J. Hyatt Hammond and Associates
(courtesy of Pella Corporation)

Many buildings today are being designed with a high degree of transparency for the enclosure. In this case, individual windows are replaced by window walls, where glass spans from floor to ceiling. Glass allows for daylighting as well as transparency between the indoors and outdoors, while the views of the outdoors have been associated with greater satisfaction with the indoor environment by the occupants. For these reasons, the use of large glass areas for entry lobbies, atriums, and occupied zones of the building such as offices has become a common design approach. Because glass typically does not have the structural strength, particularly in tension, as other common wall materials, it must include a supporting structure such as mullions. This is because glass is an amorphous solid that offers little tensile strength unless it is laminated with reinforcing sheets, such as plastic, between glass layers.

A problem with the structural mullion is that it is typically opaque and therefore counteracts the overall transparency of the glazing. To maintain a high degree of transparency ultra-thin mullions may be used or the glass may be butt-jointed with only a small seam. Unfortunately, these approaches provide little lateral resistance to wind loads. A cable truss system introduces an interior lattice of vertical tension cables with horizontal compressive bars attached to the window with spider connectors at regular spacing. The result is a visually light structural system as in the lobby of 88 Wood Street, London, designed by Richard Rogers partnership. The bow truss system is shown in Figure 4.22.

Another solution for highly transparent facades is to use vertical interior glass panels attached to the glass wall at a perpendicular angle. Attached at regular intervals, these glass fins act as small shear walls providing resistance against the force of the wind.

Figure 4.22 Bow truss system (courtesy of Thermosash Inc.)

Shear Walls – Responding to Positive Pressurization

As shown in Equation (4.1), wind exerts a pressure on the building facades that is proportional to the square of the velocity. For medium and high-rise buildings this pressure must be addressed through structural load analysis and wind load resistance strategies. Wind-driven forces that push on the wall can potentially cause failure. The magnitude of these forces on any given wall, depends, in part, on the geometry (width and height) and the unsupported length between structural elements. A commonly used structural system to respond to these lateral wind forces is to introduce shear walls. There are numerous references to shear wall design that will not be repeated here. For example, the American Iron and Steel Institute (AISI) has released the *Cold-Formed Steel Shear Wall Design Guide*, 2019 edition (AISI D113–19), which presents design guidelines and examples of steel sheet and wood structural panel sheathed, cold-formed steel framed shear wall assemblies used to resist wind and seismic forces. However, responding to wind-driven forces using shear walls can be expressed through design. [16]

Britannica defines a shear wall as,

> in building construction, a rigid vertical diaphragm capable of transferring lateral forces from exterior walls, floors, and roofs to the ground foundation in a direction parallel to their planes. Examples are the reinforced-concrete wall or vertical truss. Lateral forces caused by wind, earthquake, and uneven settlement loads, in addition to the weight of the structure and occupants, create powerful twisting (torsional) forces. These forces can literally tear (shear) a building apart. Reinforcing a frame by attaching or placing a rigid wall inside it maintains the shape of the frame and prevents rotation at the joints. Shear walls are especially important in high-rise buildings subject to lateral wind and seismic force*s*. [17]

As a response to wind forces, the shear wall can have formative influences on both the building facade and appearance, as well as the interior spatial layout. For example, exterior shear walls may be described as perforated or segmented, Figure 4.23. The perforated shear wall tends to be a structurally continuous and rigid wall with relatively small openings for windows. These small openings can limit access to daylight or views that are important for resource conservation and the health and well-being of building occupants. The segmented shear wall, Figure 4.23 (bottom), tends to divide the elevation into alternating segments of opaque/structure and transparent windows. This has consequences for both the design of the facade and the corresponding layout of interior spaces.

In their paper titled "Shear Wall Layout Optimization for Conceptual Design of Tall Buildings," Yu Zhang and Caitlin Mueller present a typology of shear wall systems, both exterior and interior, many that directly impact the layout of rooms and floor planning for high-rise buildings. [18] Many of the solutions proposed by Zhang and Mueller include interior structural walls that provide rigidity to the building frame. Because these walls transfer loads from the upper floors to the foundation, their layout and section must be continuous from floor to floor. This imposes constraints on the space planning as the shear walls cannot be altered or penetrated with large openings.

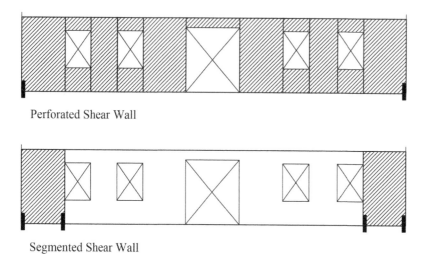

Perforated Shear Wall

Segmented Shear Wall

Figure 4.23 Perforated (top) and segmented (bottom), source: author's figure

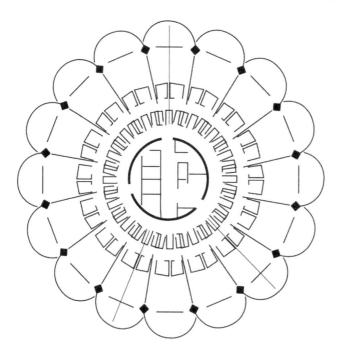

Figure 4.24 Chicago Marine City Towers by Bertrand Goldberg, source: author's figure

Shear walls may be constructed of concrete, masonry, or wood, and, sometimes in tall buildings, steel. Shear walls may be arranged to form a box where openings may be restricted in size or location or may be arranged perpendicular to the exterior walls so as to resist the lateral forces and reduce the unsupported length. Recognizing the necessity of shear walls can be translated into an aeroform opportunity.

The Chicago Marine City project, designed by Bertrand Goldberg in 1959 and completed in 1964 is another example of the integration of shear walls. The two high-rise towers each have a circular plan with a central service core, Figure 4.24. The separation of the living units is by radial partitions that, along with the core, act as shear walls. [19–20]

Tube Construction: Responding to Wind Resistance

In large part, due to the design restrictions mentioned in the previous section, the shear wall, particularly the exterior shear wall, has fallen out of favor with architects and structural engineers. Since the mid-1960s, in many cases, the shear wall has been replaced by the tube construction system. *In structural engineering, the* **tube** *is a system where, to resist lateral loads (wind, seismic, impact), a building is designed to act like a hollow cylinder, cantilevered perpendicular to the ground. This system was introduced by Fazlur Rahman Kahn in 1965 while at the architectural firm Skidmore, Owens & Merrill (SOM), in their Chicago office. The first example of the tube's use is the 43-story Khan-designed DeWitt-Chestnut Apartment Building, since renamed Plaza on DeWitt, in Chicago, Illinois, finished in 1966.* [21] The system can be built using steel, concrete, or composite construction (the discrete use of both steel and concrete). It can be used for office, apartment, and mixed use. Most buildings of over 40 stories built since the 1960s are of this structural type.

The tube structural system presents an opportunity to visually express a design response to wind forces on high-rise buildings. Luthador presents a typology of tube framing that provides a number of choices for visual expression, Figure 4.25. [22] Since its introduction, thousands of buildings, many quite notable, have been designed and built with an expressive tube structural system. The John Hancock Center in Chicago is a popular visitor attraction and icon for the Chicago cityscape, Figure 4.26. The trussed tube system, with its mesh of vertical, horizontal, and diagonal elements, gives the building its iconic elevations while expressing resistance to wind forces.

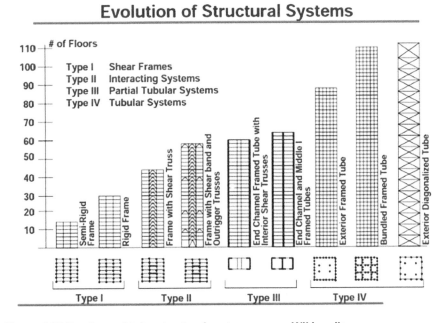

Evolution of Structural Systems

Figure 4.25 Typology of tube structural system, source: Wikimedia

Figure 4.26 John Hancock Center, Chicago Illinois by SOM, source: Wikimedia

The Sunscreen as a Pressure-Reducing Strategy

As mentioned earlier, many buildings are being designed today for a high degree of transparency, with a large window area to wall area ratio. A concern for this design intention is that glass transmits solar radiation and can also contribute to glare. A common solution to these concerns is to introduce an external sunscreen as a mask to the building and to intercept the unwanted solar rays before they pass through the window. Often, the shading elements of the sunscreen are fragile and susceptible to damage from wind forces. In this

Figure 4.27 *New York Times* **headquarters by Renzo Piano, source: Wikimedia**

case, either the shading elements are designed to be sturdy and resistant to the wind or another layer is added to the system that acts to reduce the wind forces on the shading elements. Designed by the Renzo Piano Workshop, the *New York Times* headquarters in New York City is an example of a sunscreen that also functions as a windscreen, and becomes the facade of the building, see Figure 4.27.

As described in Lawrence Berkeley Lab's Science Beat,

the use of floor-to-ceiling glass maximizes light and views for people inside and outside the building. The horizontal white ceramic rods on the building facade, which are spaced to allow occupants to have unobstructed views while both seated and standing, act as an aesthetic veil and a sun shade. They are made of aluminum silicate, an extremely dense and high-quality ceramic chosen for its durability and cost-effectiveness. Glazed with a finish similar to the material used on terracotta to reflect light, self-clean, and resist weather, the rods change color with the sun and weather. Additionally, the automated louver shades move in response to the position of the sun and inputs from sensors, blocking light to reduce glare or allowing it to enter at times of less direct sunlight. The movable shades reduce energy consumption about 13% by reducing solar heat gain by 30%. [23–25]

A variation of the sunscreen is the double envelope or double-skin glass facade. The term "double-skin" refers to an arrangement with a glass skin in front of the actual building facade. Double-skin glass facades are typically three layers of glass, one double-pane unit, and the other single-pane, separated by a ventilated air cavity, as in Figure 4.28. Solar control devices such as shades or louvers are typically placed in the cavity between the two skins. This reduces the heat gain resulting from solar radiation transmitted into the cavity by up to 25%.

One of the suggested benefits of double-skin facades is that they allow for natural ventilation by acting as a wind screen. The outer layer of the double skin reduces the wind-driven pressure for operable windows located in the inner layer. As a result, for perimeter zones, windows can be opened without significant turbulence and disturbance from the natural ventilation airflow. For medium or high-rise buildings, as the building height extends to the upper range of the surface boundary layer, wind speeds increase. If allowed to flow directly into occupied zones of the building, this fast-moving air could create conditions of draft or blustery indoor conditions. The double skin can act to reduce these conditions. This was part of the strategy for both the Commerzbank and the RWE Ag buildings, as shown schematically in Figure 4.28.

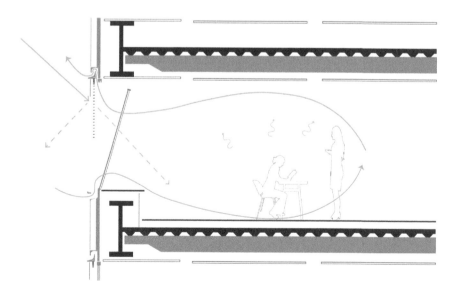

Figure 4.28 Double-skin facade system in Commerzbank building, source: author's figure

As previously introduced, the effects of wind on buildings are both static and dynamic. Oesterle et al. suggest that the double-skin facade is most effective for dampening the dynamic effects. They state that

> "the effect of sudden changes of wind pressure is dampened in double-skin facades by the combined effect of openings in the inner and outer skins in conjunction with the intermediate space, which acts as a buffer. This characteristic is particularly important in high-rise buildings, where peak wind pressures caused by gusts can be many times greater than the static pressure.
>
> On the other hand, a more or less constant wind pressure will spread through the rooms" as a result of the more stable static pressures. [26]

Similar to the double envelop system, windscreens not only significantly reduce the wind pressure on the building facade but can also act to protect windows from damaging debris during high wind events. For example, Hendee Enterprises, Inc. manufactures a variety of windscreens and hurricane protection systems. Their Force 12 Hurricane Protection system is a new 100% polypropylene basket woven fabric designed to repel flying debris for category 5 hurricanes. [27] This screen is reported to take a 120 mph wind on one side and reduce it to 3 mph on the other.

Conclusion

The understanding and response to lateral wind forces are essential if buildings are to stand and structural damages to be avoided. However, beyond these functional concerns, the siting of the building, shape, and use of pilotis, shear walls, double envelopes and windscreens offer opportunities for expression; they represent aeroform.

References

[1] Wikipedia (2022) *30 St. Mary Axe*. Available from: https://en.wikipedia.org/wiki/30_St_Mary_Axe#/media/File:30_St_Mary_Axe_from_Leadenhall_Street.jpg [Accessed October 2019].

[2] CIBSE (2005) *CIBSE Application Manual AM10: Natural Ventilation in Non-Domestic Buildings*. The Chartered Institution of Building Services Engineers London. Norwich, Norfolk Great Britain, Printed by Page Brothers Ltd., p. 55.

[3] University of Idaho (2022) *Creating Microclimates – Control of Wind*. Available from: https://webpages.uidaho.edu/larc453/pages/microclimates.htm [Accessed October 2019].

[4] Brown, G.Z. and M. DeKay (2001) *Sun, Wind and Light: Architectural Design Strategies*, 2nd edition. New York, NY, John Wiley and Sons Inc.

[5] Evans, B.H. (1957) *Natural Ventilation Around Buildings*. Texas Engineering Experiment Station, Research Report 59. College Station, TX, Texas A&M University, March.

[6] West, A. (2000) *Exploration of the Natural Ventilation Strategies at the World Trade Center Amsterdam*. Master of Science thesis. Virginia Polytechnic Institute and State University, Blacksburg, VA.

[7] Moore, F. (1993) *Environmental Control Systems – Heating, Cooling, Lighting*. New York, NY, McGraw-Hill Inc., p. 180.

[8] Bowen, A. (1981) *Classification of Air Motion Systems and Patterns*. Proceeding of the International Passive and Hybrid Cooling Conference, American Solar Energy Society, Boulder, CO, pp. 743–763.

[9] ASHRAE (2005) *ASHRAE Fundamentals – F16: Airflow Around Buildings*. The American Society of Heating, Refrigerating and Air-conditioning Engineers. From Swami, M.V. and S. Chandra. 1987. *Procedures for*

Calculating Natural Ventilation Airflow Rates in Buildings. Final Report FSEC-CR-163-86. Cape Canaveral, Florida Solar Energy Center.

[10] KPF (2016) Email from Kohn Pedersen Fox: Literature and Project Description Provided to James Jones, September.

[11] Blake, P. (1996) *Master Builders*. New York, NY, W.W. Norton & Company, pp. 57–59.

[12] Jencks, C. (1976) *LeCorbusier and the Tragic View of Architecture*. Cambridge, MA, Harvard University Press, pp. 110–111.

[13] IBC (2006) *ICC International Building Code*. Washington, DC, International Code Council.

[14] ASCE (1998) *Minimum Design Loads Buildings and Other Structures – Standard 7–1998*. New York, NY, American Society of Civil Engineers.

[15] Novak, C.A. (2008) *Targets for Building Performance: Selecting Windows That Work. Architectural Record*. New York, NY, McGraw-Hill Publishing, pp. 251–254.

[16] American Iron and Steel Institute (AISI) (2019) *Cold-Formed Steel Framed Wood Panel or Steel Sheet Sheathed Shear Wall Assemblies*. Washington, DC, American Iron and Steel Institute.

[17] Britannica (2022) *Shear Wall*. Available from: www.britannica.com/technology/shear-wall [Accessed October 2019].

[18] Zhang, Y. and C. Mueller (2017) "Shear Wall Layout Optimization for Conceptual Design of Tall Buildings," *Engineering Structures* 140: 225–240. Available from: https://doi.org/10.1016/j.engstruct.2017.02.059 [Accessed October 2019].

[19] Goldberg, B. (2022) *Marine City*. Available from: http://bertrandgoldberg.org/projects/marina-city/ [Accessed October 2019].

[20] Wikipedia (2022) *Marine City*. Available from: https://en.wikipedia.org/wiki/Marina_City [Accessed October 2019].

[21] Wikipedia (2022) *Tube Structural System*. Available from: www.designingbuildings.co.uk/wiki/Tube_structural_system [Accessed October 2019].

[22] Wikipedia (2022) *Tube Structural System*. Available from: https://upload.wikimedia.org/wikipedia/commons/f/f4/Skyscraper_structure.png; Luthador, CC BY-SA 3.0 <https://creativecommons.org/licenses/by-sa/3.0>, via Wikimedia Commons [Accessed November 2019].

[23] Science Beat (2022) "Science Beat: Berkeley Lab: The New York Times Building," *Designing for Energy Efficiency Through Daylighting Research*. Available from: https://www2.lbl.gov/Science-Articles/Archive/sb-EETD-NYT-building.html [Accessed November 2019].

[24] Science Beat (2004) "The New York Times Building: Designing for Energy Efficiency Through Daylighting Research," in *Science Beat*. Berkeley, CA, University of California, February 17.

[25] Jambhekar, S. (2004) *Times Square Skyscrapers: Sustainability Reaching New Heights* (PDF). CTBUH Conference, Seoul, October 10.

[26] Oesterle, E., R.D. Lieb, M. Lutz and W. Heusler (2001) *Double-Skin Facades*. Munich, Prestel Verlag.

[27] Hendee (2010) *Force 12 Hurricane Protection*. Available from: www.hendee.com/products/force_12_hurricane_protection [Accessed November 2019].

Chapter 5

Roofs and Response to Uplift

According to the National Weather Service, the 2005 Atlantic hurricane season was the most active in the 154 years that records have been kept. In that year, there were 27 named storms, with three reaching category 5. Hurricanes Katrina ($80 billion) Figure 5.1, Rita ($9.4 billion), and Wilma ($14.4 billion) together accounted for over $103.8 billion in damage costs, with Katrina being the costliest storm in US history. [1] Because of the apparent increasing frequency of severe storms and the magnitude of damage, architects must understand and design buildings for these high wind events and their impact on architectural form.

In the United States, buildings must be designed to resist both lateral and uplifting wind forces. The resistance of these forces is of particular importance for high wind and at-risk locations, such as ridge tops in mountainous zones and coastal locations in hurricane-prone regions. For these situations, construction requirements for uplift resistance can be quite prescriptive. Florida and other coastal areas in the southeast have building codes that require roofs to be designed and installed to resist wind speeds of 100 to 150 mph. This often influences construction techniques, such as the introduction of strapping approaches that tie the roof to the wall and/or structure of the building, as will be shown later.

DOI: 10.4324/9781003167761-5

Figure 5.1 Satellite image of Hurricane Katrina, source: Wikimedia

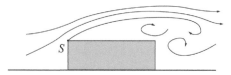

Figure 5.2 Wind flow pattern over a rectangular building, source: author's figure

Reducing Uplift through Roof Geometry

One of the iconic images of hurricane damage is the roof lifting from the walls of a building and flying away. As introduced in Equation (3.5) in Chapter 3, the Bernoulli equation suggests that as airflow accelerates, pressure is reduced relative to the ambient pressure. As wind encounters a building, the flow streams are compressed over the building as in Figure 5.2, and speed increases, inducing lower pressure and uplift on the roof. The roof, in effect, acts as an airfoil and attempts to "take off" from the walls of the building. The Bernoulli equation can be adapted to Equation (5.1), which estimates the negative (uplift) pressure (w) on the roof. Under these conditions, the internal and external pressures generated on a structure can be greatly influenced by elevation, building shape, openings, and surrounding terrain. Understanding these influencing factors can lead to an aeroform response.

$$p - p_{\infty} = C_{\rho} V_{\infty}^{2} / 2 \qquad (5.1)$$

Where p_{∞} is the ambient pressure, ρ is the density of air (approximately 0.0026 slugs/ft^3), V_{∞} is the wind velocity upstream of the structure in ft/s, and C is a roof pressure coefficient

Substituting an estimate for C = 1 for high wind conditions, we get Equation (5.2).

$$p - p_{\infty} = .0.0026 \, V_{\infty}^{2} / 2 \qquad (5.2)$$

For a 100 ft/s (68 mph) wind, the uplifting pressure would be about 13 lb/ft^2, which is a reason for high wind failure.

The surface pressure coefficient on a sloped roof depends in part on the roof slope angle. In Figure 5.3 from the 2013 ASHRAE Fundamentals chapter "Airflow Around Buildings," [2] the length of the arrows indicate the magnitude of the uplifting surface pressure coefficient. As shown, for flat and low-sloped roofs (less than 15°), the pressure coefficients are lowest (most negative) on the windward half of the roof and uplift is greatest at the windward edge of the roof (scale is given to the right of top two images). It is here where failure is most likely to initiate, and response strategies such as structural strapping should be applied. For greater sloping angles, the dominant low-pressure field moves to the leeward roof surface. For the 15° slope, the lowest pressure occurs near the roof ridge, while for the 20° and 30° slopes, the lowest pressure occurs at the leeward edge of the roof. What may be most significant is that for slope angles of 15° and higher, the overall uplifting forces are smaller when compared to the low slope roofs, suggesting that higher sloped roofs may be preferred to low-sloped solutions if reducing the uplift is a goal (Figure 5.3).

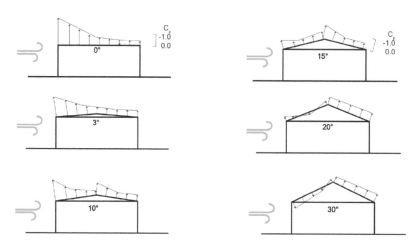

Figure 5.3 Uplift due to wind for alternative roof slope angles [5], source: author's figure

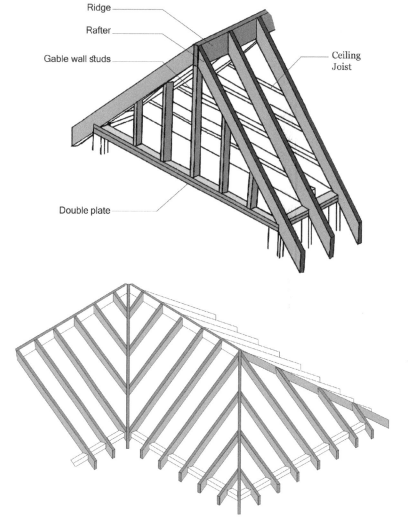

Figure 5.4 Gable and Hip roof construction, source: author's figure

Roof geometry and construction may also play a role in reducing uplift. For example, past studies have suggested that hip roofs survive high wind conditions better than gable roofs. While it may seem that geometric differences in these roofs result in different aerodynamic behavior, D. Meecham [3] suggests that the overall lift and overturning loads on these two roofs are almost the same. He goes on to suggest that a higher survival rate for hip roofs is in large part due to the difference in structural framing of hip roof trusses, which experience lower structural stress than gable trusses due to their construction geometry (Figure 5.4). In particular, trusses in gable roof systems are typically not laterally braced other than by the connections to the subroof. Without this lateral bracing, gable roof systems often shift and tilt, and finally collapse. Constructing a gable roof system with horizontal bracing between the trusses can significantly improve the resistance to wind damage. According to a home construction guide from the National Oceanic and Atmospheric Administration, the truss bracing usually consists of 2 x 4s that run the length of the roof. If more than one length is needed, they should overlap on the ends, and the overlap should extend across at least two trusses. Braces should be installed 18 inches from the ridge in the center span and at the base, with 8 to 10 ft between braces. Bracing strategies are shown in the *Bracing of Roofs for Hurricane, High Wind, and Seismic Resistance Guide* from the Building American Solutions Center website and in Figure 5.5. Use two 3-inch, 14-gauge wood screws or two 16d galvanized common nails to attach the 2 x 4s to the gable and to each of the four trusses. [4]

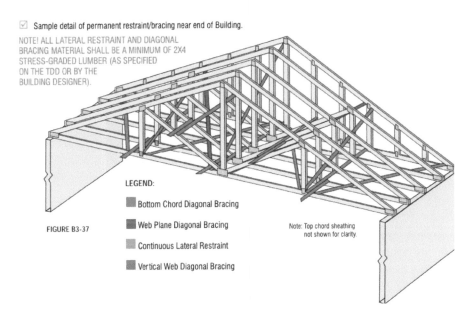

☑ Sample detail of permanent restraint/bracing near end of Building.

NOTE! ALL LATERAL RESTRAINT AND DIAGONAL BRACING MATERIAL SHALL BE A MINIMUM OF 2X4 STRESS-GRADED LUMBER (AS SPECIFIED ON THE TDD OR BY THE BUILDING DESIGNER).

FIGURE B3-37

LEGEND:

■ Bottom Chord Diagonal Bracing

■ Web Plane Diagonal Bracing

■ Continuous Lateral Restraint

■ Vertical Web Diagonal Bracing

Note: Top chord sheathing not shown for clarity.

Figure 5.5 Alternative truss bracing for high wind resistance, source: author's figure

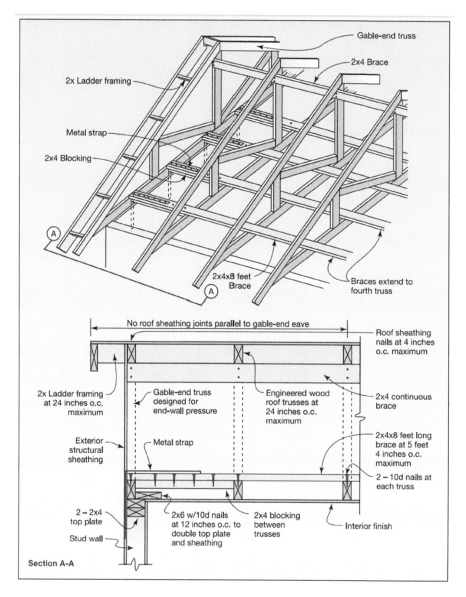

Figure 5.5 Continued

Roof Vent Strategies

A common weatherproofing layer for flat or low-sloped roofs is membranes. Typically composed of materials such as rubber or polyvinylchloride, these roofing products are gaining popularity because of their high solar reflective properties that reduce the roof sol-air temperature and the impact of the sun on space cooling loads. A concern for these roofing systems is that they tend to lift and tear during high wind events. Solutions to this problem include physically attaching the membrane to the subroof, fully adhering (gluing) to the roof, or ballasting the roof with stone or concrete pavers, all of which add cost to the system and potentially reduce thermal performance.

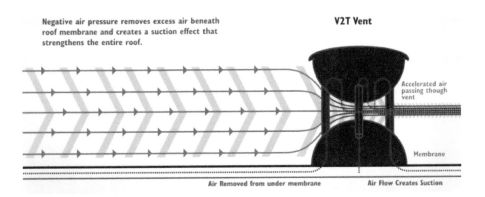

Negative air pressure removes excess air beneath
roof membrane and creates a suction effect that
strengthens the entire roof.

V2T Vent

Accelerated air
passing though
vent

Membrane

Air Removed from under membrane   Air Flow Creates Suction

Figure 5.6 V2T roof vent system (courtesy V2T Roof systems Inc.)

A recently developed omnidirectional roof vent has the potential to counteract the uplifting forces on the roof while reducing installation costs (Figure 5.6). As shown in the image, this vent takes advantage of the Bernoulli effect where air is directed between two surfaces and accelerates to produce a low-pressure region near the point of least separation between upper and lower dome elements. By coupling this low-pressure region to the underside of the membrane through hollow vertical supports, the pressure under the membrane becomes low relative to the ambient pressures and the uplift on the membrane is relieved. In fact, the vent acts as a vacuum to draw the membrane down onto the subroof. Full-scale tests in the high-speed wind tunnel at Virginia Tech indicated that this new vent could produce lower than -1.11 Cp suction at wind speeds of near 150 mph. [6] These same principles can be applied to induce airflow for natural ventilation, as will be discussed later.

Straps and Tie-Downs

One strategy for holding the roof in place during high wind events is to use straps and tie-downs to attach the roof structure to the primary load-carrying walls or to the foundation. By securing the roof to walls, the uplifting forces can be transferred to the building's foundation. Most homes built in Florida since 1995 have some type of tie-down connections, such as galvanized metal roof clips or straps that tie the roof to the exterior walls. Hurricane straps are available from a number of different vendors and are usually of a size and metal thickness for this purpose. The straps must be anchored to the roof system and walls with either screws or a gusset plate as shown in Figures 5.7, 5.8, and 5.9.

It has been suggested that architecture is the art of building where exposing and expressing methods of construction can often be an opportunity to demonstrate this position. Constructing the building in response to uplift through the use of straps and ties might be viewed as an expressive opportunity. Here the transfer of structural forces and resistance to uplift can be made visible by revealing the connections rather than hiding them.

Figure 5.7 Schematic hurricane strap installation, source: Wikimedia

Figure 5.8 Image of structural ties, source: Wikimedia

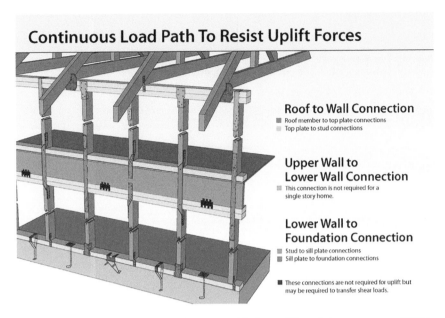

Figure 5.9 Schematic of construction assembly for uplift resistance, source: IIHBS

References

[1] infoplease (2010) *Billion Dollar U.S. Weather-Disasters 1980–2009*. Available from: www.infoplease.com/ipa/A0882823.html [Accessed November 2019].

[2] Holmes, J.D. (1983) *Wind Loads in Low-Rise Buildings – A Review*. Commonwealth Scientific and Industrial Research Organization (CSIRO). Division of Building Research Australia. As cited in: ASHRAE (2013) *ASHRAE Fundamentals Chapter F24: Airflow Around Buildings*. Atlanta, GA, The American Society of Heating, Refrigerating and Air-Conditioning Engineers.

[3] Meecham, D. (1992) "The Improved Performance of Hip Roofs in Extreme Winds – A Case Study," *Journal of Wind Engineering and Industrial Aerodynamics* 43(1–3): 1717–1726. Tokyo, Japan, published by the International Association of Wind Engineering.

[4] SBCCI (1999) *Southern Building Code Congress International Document SSTD 10–99: Guidelines for New Home Construction in Hurricane-Prone Areas*. Washington, DC, The International Code Council.

[5] Holmes, J.D. (1983) *Wind Loads on Low-Rise Buildings – A Review: Commonwealth Scientific and Industrial Research Organization (CSIRO)*. Reston, VA, Division of Building Research, Australia.

[6] Grant, E. (2003) *Design and Implementation of a Pressure-Equalizing Vent System for Low Slope Roofs*. Master of Science thesis, The College of Architecture and Urban Studies, Virginia Tech, Blacksburg, VA.

Chapter 6

Natural Ventilation

One of the most expressive opportunities for aeroform is the integration of natural ventilation. Natural ventilation can have the twofold benefit of providing a positive psychological connection between inside and out for the building occupants, while potentially reducing the need for mechanical cooling and forced ventilation. Because of this, there is a growing interest from architects and architectural engineers in the subject of natural ventilation.

The application of natural ventilation is not an exclusively human strategy. Termite mounds in hot climates such as in Africa and South America are often built with openings placed both near the bottom and top of the structures. These openings allow airflow through the interior cavity of the mound providing thermal regulation for the termite larvae.

The integration of natural ventilation into buildings is not a contemporary strategy. Indeed, there are many historical examples of buildings designed with unforced airflow in mind. The tipi structures of Native American Indians of the Great Plains, for example, were designed to be easily transported while the openings at the top and around the bottom perimeter allowed for natural air movement that provided a cooling breeze and exhausted smoke from the interior of the structure.

The courtyard houses from the southwest United States and the Middle East take advantage of warm days and cool nights and thermal buoyancy conditions to induce air movement through the building (Figure 6.1). The malquafs towers in the Middle East (Figure 6.2) have been used for over 500 years to both catch the wind and exhaust air to provide cooling to the building inhabitants. And the layout of George Washington's Mt. Vernon was designed to capture prevailing breezes and pass the air through subterranean passages before being introduced to the occupied zones of the house. These are but a few historical examples of design to enhance natural ventilation.

DOI: 10.4324/9781003167761-6

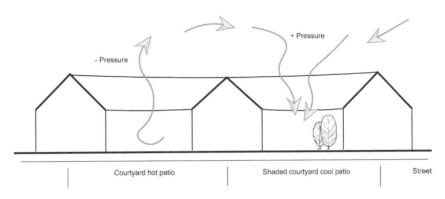

Figure 6.1 Schematic natural ventilation through courtyard design, source: author's figure

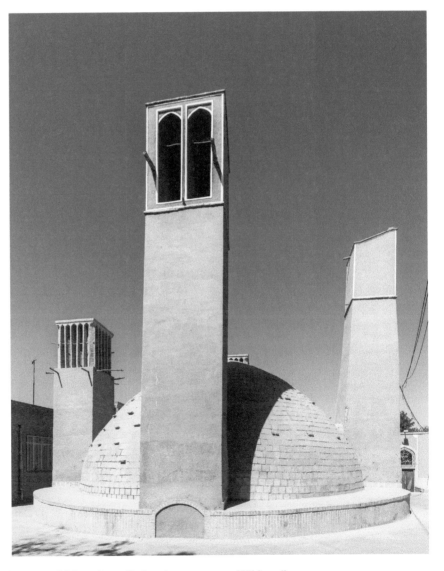

Figure 6.2 Malquaf ventilation tower, source: Wikimedia

To begin the process of designing for natural ventilation, one should consider the appropriateness of the strategy for the intended function of the building. Some building types such as laboratories, where tight zone-to-zone pressure control is required, may not lend themselves to natural ventilation. Likewise, libraries that often need tight temperature and humidity control to prevent degradation of the book bindings may not be appropriate for natural ventilation in humid locations. The operation of the building can impose constraints and limits on the integration of natural ventilation that should be considered as part of the design process.

In her dissertation a *Decision-Support Framework for the Feasibility of Natural Ventilation of Non-residential Buildings*, Ying Zhao develops a decision-support framework that includes these and other design decision-making factors. [1] She suggests that in addition to the intangible benefits of providing a connection between inside and out, there are at least five tangible performance factors that should be considered; these include relative energy savings, thermal comfort, indoor air quality, acoustics, and the sonic environment and cost. While too lengthy to fully present here, Zhao describes a process for integrating natural ventilation into nonresidential buildings that include suggested performance assessment algorithms and interactive considerations.

Zhao suggests that another consideration for natural ventilation is whether the performance intent is, at a general scale, (a) to reduce energy consumption or (b) to provide an occupant connection between inside and out, and at a more specific scale, to (1) remove heat, (2) provide occupant cooling, or (3) both. While these intentions may be interrelated, priority of one over the others may inform the design process, as will be shown later.

Benefits of Natural Ventilation: Reducing Energy Consumption

There are at least two important benefits to the integration of natural ventilation in buildings. First natural ventilation can reduce energy consumption. The operation of heating, ventilating, and air-conditioning systems has two primary functions: (1) to maintain acceptable thermal control in the space and (2) to provide outdoor air ventilation for indoor air quality control. Both of these functions could be met with natural ventilation. For many temperate climate regions of the world, including many coastal and southern zones in Europe and the central United States, there are many hours during the year when outdoor air conditions are in or near the thermal comfort zone and outdoor air could be introduced with little if any mechanical conditioning. Even extreme climate zones such as hot, arid regions can benefit from natural ventilation by employing a day (closed) – night (open) strategy. Computer-based tools such as Climate Consultant can be applied to analyze outdoor air conditions as compared to the thermal comfort zone and estimate the annual hours of operation when natural ventilation could be implemented. Figure 6.3 shows the annual percent savings in cooling energy possible from good natural ventilation in a low-mass residence. [2] As will be discussed later, these percentages are likely higher for nonresidential buildings as these tend to be internal load-dominated buildings with lower balance points and therefore more hours per year when outdoor conditions could be used to cool the building. Note that the balance point *of a building is the outdoor air temperature at which the rate of internal heat gain equals the rate of heat loss through the building enclosure*

Figure 6.3 Annual cooling energy reduction from natural ventilation in low-mass residential construction [4], Source: S. Chandra, P. Fairey, M. Houston

to the outdoors. Buildings with more internal heat gain tend to be nonresidential and have lower balance points. Also, nonresidential buildings tend to be constructed of higher thermal mass materials such as concrete that can be utilized to absorb the cooling effects of outdoor air, particularly at night, and act as a thermal flywheel during early hours of operation to reduce mechanical cooling.

Benefits of Natural Ventilation: Occupant Satisfaction

The second benefit from natural ventilation is greater occupant satisfaction with the indoor environment. In his paper "User and Occupant Controls in Office Buildings," W. Bordass shows that the level of perceived comfort in a sample of naturally ventilated buildings was similar to that in a group of air-conditioned buildings, and the level of perceived control, particularly for ventilation, was considerably higher for the naturally ventilated buildings. [3] This would seem to imply a higher level of satisfaction with the naturally ventilated building, as the occupant has the choice to open or close a window for thermal comfort, indoor air quality control, or to open the space to the outdoor air.

In her book *Thermal Delight in Architecture*, Lisa Heschong [5] suggests, "*People have a sense of warmth and coolness, a thermal sense like sight or smell, although it is not normally counted in the traditional list of our five senses.*" She goes on to say,

> There is a basic difference, however, between our thermal sense and all of our other senses. When our thermal sensors tell us an object is cold, that object is already making us colder. If, on the other hand, I look at a red object it won't make me grow redder, nor will touching a bumpy object make me bumpy.

She goes on to suggest that the environments that surround our lives provide clues with regard to thermal sensations.

> Such clues from other senses can become so strongly associated with a sense of coolness or warmth that they can occasionally substitute for the thermal experience itself. For example, the taste of mint in a drink or food seems refreshing and cool regardless of what temperature it is . . . .. Many of the other sensory associations with cooling seem to want to remind us of something, like the breeze, lightly playing over a surface.

And so as introduced in Chapter 1, the natural ventilation scheme used in the Bluewater Mall project (Figure 6.4) was, in part, intended to reintroduce the variation in air movement experienced along a European street back into the shopping experience, and the alternating openings in Mario Botta's chapel at Monte Tamaro, through air movement, provide a rhythm as one moves along the approach to the chapel. These strategies go beyond saving energy to providing a psychological connection between inside and out.

Similarly, the Commerzbank building by Norman Foster and Partners employs a compact plan with daylight and natural ventilation as an architectural response to the corporate transparency that is at the philosophical core of the Commerzbank's operations. The building skin is not only transparent to light but windows can be opened by the occupants to allow a cooling, refreshing breeze to flow through the office zones. The result is a higher level of control and satisfaction, as implied in the Bordass study.

Figure 6.4 Bluewater Mall roof ventilators (courtesy Vision Ventilation Co. UK)

Thermal Comfort and Natural Ventilation

ASHRAE Standard 55, *Thermal Environmental Conditions for Human Occupancy* [6] states that "*thermal comfort is a state of mind that expresses satisfaction with the thermal conditions in a space.*" The standard suggests that thermal comfort is primarily the result of six factors: activity and metabolic rate, insulating value of clothing, air temperature, humidity, mean radiant temperature, and air speed. At least three of these factors are relevant for natural ventilation: air temperature, humidity, and air speed.

Until recently one of the constraints to the implementation of natural ventilation in nonresidential buildings was indoor environmental quality standards such as ASHRAE 55 and 62. In their 2000 paper titled "A Standard for Natural Ventilation," Gail Schiller Brager and Richard de Dear laid the groundwork for a more flexible thermal comfort standard, arguing "*that the primary limitation of Standard 55 is its 'one-size-fits-all' approach, where clothing and activity are the only modifications one can make to reflect seasonal differences in occupant requirements,*" were too limiting. [7] They say "*that the heat-balance model of thermal comfort underlying the present standard cannot account for the complex ways people interact with their environments, modifying their behaviors, or gradually adapt their expectations to match their surroundings.*" As a result of this research, ASHRAE has now developed an adaptive thermal comfort model in which issues such as adaptive control (opening windows) and personal preferences can be better accounted for. The result is an extension of the previous thermal comfort zone temperature limits from about 20°C (68°F) lower to 27°C (82°F) upper limit.

In his paper titled "Acceptable Temperature Ranges in Naturally Ventilated and Air-Conditioned Offices," Nigel Oseland shows that "*the overall (temperature) range in naturally ventilated offices in winter and summer, 4.9°C and 3.9°C (8.8°F and 7.0°F), respectively, was wider than that found in air-conditioned offices, 2.6°C and 2.4°C (4.7°F and 4.3°F), respectively.*" [8] This further supports the adaptive nature of thermal response for occupants in naturally ventilated buildings.

This adaptive thermal response has implications for design assessment. As described by Zhao, to begin to assess the benefits of natural ventilation for a building in a given location, one could overlay the adapted thermal comfort zone as displayed on a psychrometric chart with hourly outdoor air temperatures from a Typical Meteorological Year file, as shown in Figure 6.5, to determine the number of hours per year when natural ventilation could be applied. By using this graphical approach with the extended adaptive thermal comfort model, more hours per year are available for the application of natural ventilation. This is the approach used in Climate Consultant.

As mentioned earlier and presented in ASHRAE Standard 55 thermal, comfort depends on at least six factors, including airflow rate. Airflow rate is probably the most significant factor for natural ventilation, as the movement of air across the skin of the occupant provides a direct cooling effect while also producing an often desirable psychological connection between indoors and out. ASHRAE Standard 55 shows that for thermal comfort, higher air temperatures can be offset with higher airflow rates over the body. For example, Figure 6.6 shows that for equal average room surface temperature (t_r) and room air temperature (t_a), i.e. ($t_r - t = 0$), a 6°F rise in temperature above the comfort

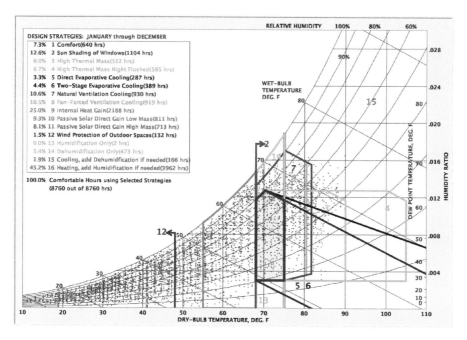

Figure 6.5 Climate Consultant output showing psychrometric and thermal comfort conditions, source: author's figure from Climate Consultant output

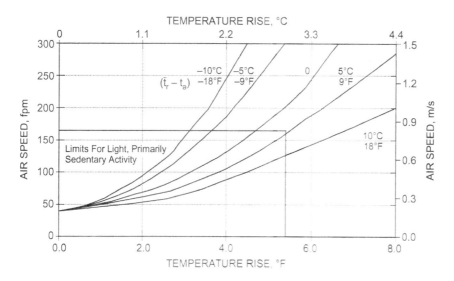

Figure 6.6 Relation of increasing temperature and airflow rate to maintain thermal comfort [9], source: ASHRAE

zone limit would require an air speed of 250 fpm to bring the occupants back into a feeling of comfort. This is one of the potential direct applications of natural ventilation, as wind through an open window will often induce airflow rates greater than those common for air-conditioned spaces.

Designing to Induce Pressure Differences

When designing buildings for natural ventilation, it is important to approach the problem in terms of pressure differences rather than only "catching" the wind. This is because airflow depends also on pressure differences within the interior of a structure. The goal is typically to create maximum or controlled pressure differences between inlets and outlets. It is also instructive to realize that there are two pressure-driving mechanisms, wind, and thermal buoyancy, and when thermal buoyancy is properly employed, natural ventilation can be achieved with little or no wind.

Wind Utilization – Site Integration

An early step in designing for natural ventilation is to site the building to capture prevailing breezes during conditions when natural ventilation is appropriate. As previously introduced, nearby landforms, buildings, or vegetation can either deflect and protect the building from winds (Figure 6.7) or act to channel the desirable breezes toward the building. Applying the psychrometric-thermal comfort zone overlay procedure and the wind rose analyses (for the months when natural ventilation is most feasible) shown previously, we can identify the directions for the desirable breezes. Then the building could be sited to take advantage of preexisting channeling conditions, or new vegetation could be planted to funnel the cooling breezes toward the proposed building.

In some cases, a controlled opening in a line of obstructions such as a row of trees can actually increase the wind speed, thus potentially providing a greater cooling effect, as in Figure 6.8. Also, for hot, arid locations, summer breezes might be channeled across a swimming pool or nearby body of water to lower the ambient temperature due to evaporative cooling and extend the natural ventilation cooling effects (Figure 6.9). On the

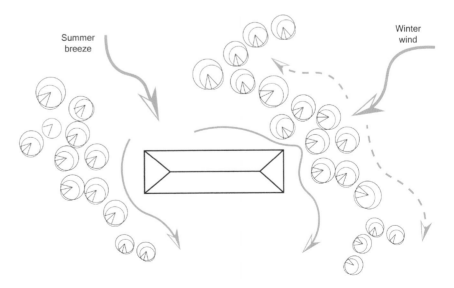

Figure 6.7 Utilization of vegetation to manage wind flow, source: author's figure

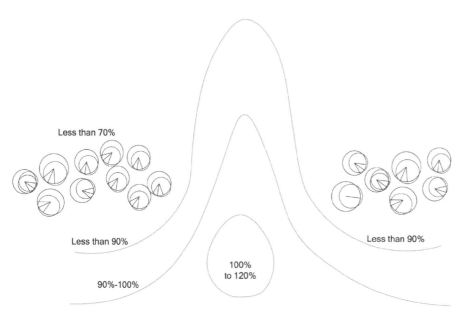

Figure 6.8 Percentage effect of opening in vegetation on wind speed, source: author's figure

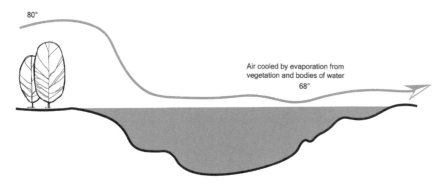

Figure 6.9 Utilizing airflow over a body of water to achieve evaporative cooling, source: author's figure

other hand, care should be exercised to not introduce ventilation air that passes over hot surfaces such as asphalt parking lots. Parking surfaces such as asphalt can be 40°F (22 °C) or hotter than adjacent grass surfaces and therefore can significantly affect the ventilation strategy.

Taking Advantage of Wind-Induced Pressures

As previously introduced, wind exerts a pressure on building enclosures that is proportional to the square of the velocity. When considering wind-induced pressurization and the resulting indoor airflow rates the relationship of the windward inlets to the leeward outlets is important. For example, as shown in Brown and DeKay, when the inlet area is smaller than the outlet area the indoor air velocity tends to be higher, thus potentially providing a more effective cooling condition. On the other hand, if maximizing the ventilation rate and air exchanges for heat removal or indoor air quality is the goal, then equal sized inlets and outlets should be the design strategy, as in Figure 6.10. The ventilation strategy should be designed with functional requirements in mind.

Another issue concerning placement and sizing of inlets and outlets is the pattern of air movement through the space. Generally, the airflow pattern through an interior space is more dependent on the position of the inlets than the outlets. For example, generally, when indoor and outdoor temperatures are similar when the inlets are positioned low and the outlets high (in section) the flow will tend to remain low in the space. If, on the other hand, the inlets are placed high and the outlets low, the airflow will tend to remain high. This has consequences for achieving the desired effect. If the design goal is to maximize the cooling effect for occupants, then low inlets will tend to direct the air through the occupied zone of the space, thus providing direct air movement across the body. This can also be effective for removing pollutants and providing fresh air to the breathing zone: between 6 inches and 6 ft in height. If the goal is to provide heat removal, because heat rises, then placing the inlets high might provide a more direct flow path for removing heat without the potentially disruptive flow of air in the occupied zone. This is shown in Figure 6.11.

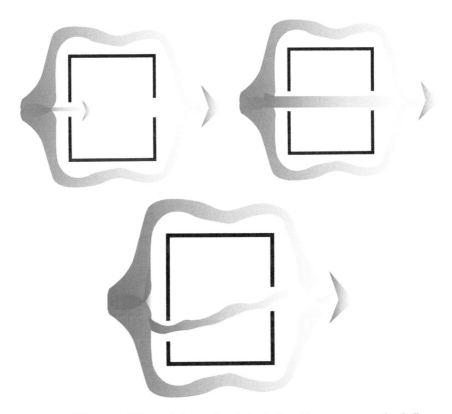

Figure 6.10 Effects of different inlet and outlet relationships, source: author's figure

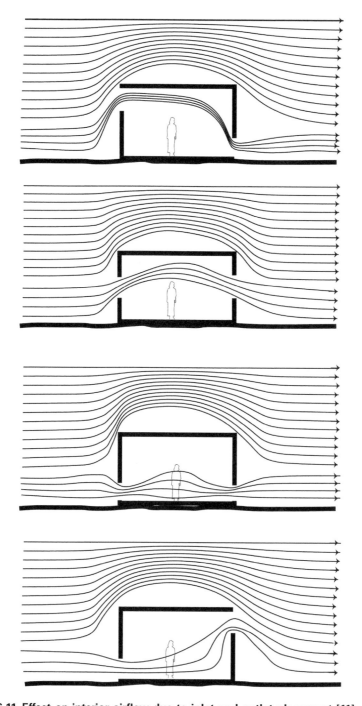

Figure 6.11 Effect on interior airflow due to inlet and outlet placement [11], source: author's figure

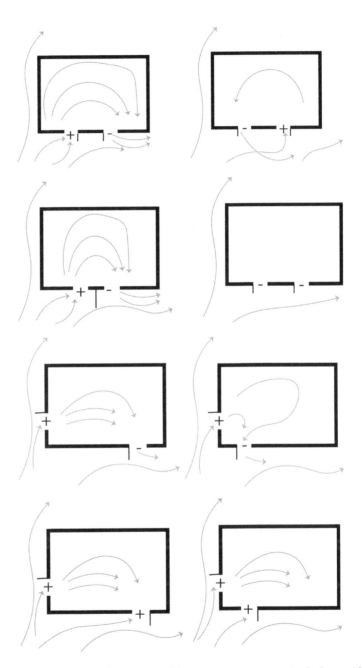

Figure 6.12 Pressure conditions created by an open casement window acting as a wingwall [12], source: author's figure

As previously stated, it is more constructive to design for natural ventilation in terms of controlling pressure differences rather than only as a response to wind direction. For example, wingwalls or the open sash of a casement window can act to either catch and divert winds into the building, producing positive pressure, or can act as a windshield, blocking the wind and producing a low-pressure zone behind the obstruction. Figure 6.12 shows alternative effects for various locations and sash opening directions for a series of rooms all subjected to the same wind direction. As shown, the conditions can be manipulated to produce maximum pressure differences (both positive and negative pressures relative to the ambient) from inlet to outlet.

Taking the Shape of the Wind

The shape of the building can also contribute to the ventilation effectiveness and desirable pressure differentials. From the Bernoulli equation, as an air stream smoothly converges and accelerates, pressure decreases in proportion to the square of the increasing velocity. This principle can be applied to whole building design by introducing a smoothly curved or domed roof. If an opening is introduced at the apex of the dome, this will act as a vacuum, drawing stale air out of the building. The Pantheon in Rome is an example of this design approach as in Figure 6.13. The same may be true for air smoothly converging through a wind tower to draw air up and out of the occupied zones as in Figure 6.14.

Similarly, as shown in Chapter 5, a low-sloped roof will act to converge and accelerate wind flowing over the building, inducing a low-pressure zone near the top of the roof. If an operable clerestory window is located in this low-pressure region, then indoor air can be drawn out of the building. For the design of a school in Blacksburg, Virginia,

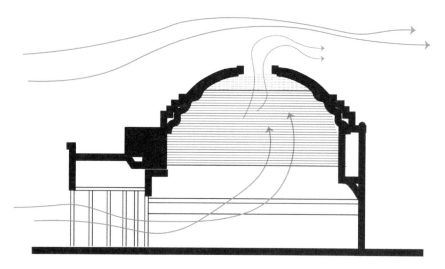

Figure 6.13 Schematic section of Pantheon showing natural ventilation. From the "Natural Ventilation" chapter by Benjamin Evans, FAIA, in Time-Saver Standards for Architectural Design Data, source: author's figure

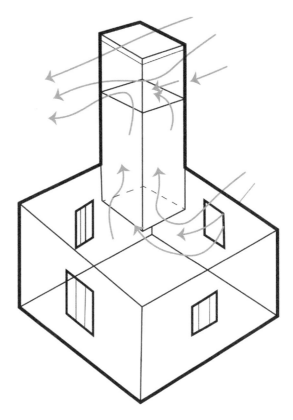

Figure 6.14 Convex opening at top of a wind tower converges the wind flow and induces low pressure to draw air out of the building, source: author's figure

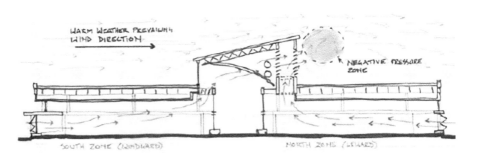

Figure 6.15 Schematic design for natural ventilation, source: J. Henderson

shown schematically in Figure 6.15, James Henderson [10] used scale models to study the roof section to achieve low-pressure outlet conditions and building natural ventilation, as in Figure 6.16.

Interestingly, when designing to create pressure differences between the inlets (left) and outlets (right), the wind tunnel study showed that in the leeward classroom (right), the airflow pattern through the classroom was actually in the opposite direction of the prevailing wind; right to left as opposed to left to right, Figure 6.16.

Figure 6.16 Wind tunnel study for natural ventilation; note: wind flow is left to right – smoke flow in classroom shown is right to left

The Shape of Hall 26

An elegant example of aeroform and shape response is Thomas Herzog's Hall 26 in Hanover Germany. An exhibit space of 25,400 m² (273,410 ft²), this building incorporates three 70-m (230-ft) bays with service cores at either end of each bay. As described by David Lloyd Jones in *Architecture and the* Environment, [13]

> The space relies for comfort on displacement ventilation. This air, however, is not introduced from the floor but from large glazed ducts that run along the line of the supporting structure, 4.1 meters (13 ½ feet) from the floor. The glazed walls of the duct ensure a sense of spatial continuity. The fresh air flows downwards distributing itself evenly over the floor. The air is then borne slowly upwards by the heat generated within the space itself and expelled through continuous openings at the ridge of each zone. The openings are controlled by adjustable flaps. For heating purposes, the ventilation system can be switched to a mode whereby pre-heated air is injected horizontally via adjustable long-range nozzles. A portion of the heated air can be recirculated.

Hall 26 integrates three wave-form roof sections that smoothly converges the prevailing winds and create a low-pressure zone at the crest of the waves, as shown in Figure 6.17 and 6.18. This low-pressure condition is enhanced by the introduction of wing-like airfoil elements just above the roof openings, as shown in Figure 6.18. These elements create a Venturi effect where the wind stream flowing over the roof converges between the top of the roof and under the airfoil, further reducing the pressure near the ridge opening. With the Venturi effect, a fluid (wind) flows through a constricted area, resulting in increased velocity and decreased static pressure. In Hall 26, the result is a very low-pressure zone that draws air up and out of the building, while at the same time, the building shape becomes a metaphor for the wave-like flow of wind.

Figure 6.17 Hall 26, Hanover Germany by Thomas Herzog (photo by Deiter Leistner)

Figure 6.18 Hall 26 roof with airfoil mounted above the ridge (photo by Dieter Leistner)

Applying the Venturi Effect to Shape the Building

The Venturi effect has been used also in a number of recently designed naturally ventilated buildings. The Ionica building in Cambridge, United Kingdom, by the RH Partnership, [14] for example, is the headquarters for a telecommunications company that incorporates a series of inverted pyramids supported above air shafts over the interior atrium to converge the wind and induce airflow out of the building, as shown in Figure 6.19. The system relies on mixed-mode operation in conjunction with the heating, ventilating, and air-conditioning (HVAC) system for achieving thermal comfort with minimal energy input.

The RWE Ag building in Essen Germany by Ingenhoven, Overdiek, and Partners is a 31-story cylindrical tower with a double-skin facade (Figures 4.27 and 6.20). As described by Klaus Daniels in his book *The Technology of Ecological Building*, [15]

> When the windows are opened, pressures from oncoming wind on the outer envelope are carried forward into the interior of the building and can create an intense cross-ventilation, depending upon interior resistance – especially doors – with such effect as – excessive force needed to open doors, drafts in office and corridors – whistling at cracks.

These limits define the extent to which natural ventilation is possible. One of the key aeroform elements of this building is the airfoil disk that acts as a hat to the building. Elevated above the top floor of the building, this convex disk compresses the wind flowing over the building, thus creating a low-pressure region above the center ventilation shaft. This element is essential for raising the neutral pressure level to the top of the building, as will be discussed in detail later.

Figure 6.19 Innovation Centre within St. John's Innovation Park in Cambridge by RH Partnership, photo by Tim Soar

Figure 6.20 Cylindrical-shaped RWE Building, Essen Germany by Ingenhoven, Overdiek, and Partners, source: Wikimedia

An important design concern for these ventilation systems is how to control rainwater that may enter through the skyward openings. For the Blacksburg school (Figures 6.15 and 6.16), Ionica (Figure 6.19), and Hall 26 building (Figures 6.17 and 6.18), the solution was to introduce a tray and drainage system under the openings with airflow from the building interior entering these rain-catching elements through louvered openings in the side rather than the bottom, thus providing a path for air to flow out while preventing rain from entering the occupied areas of the building. The system for Hall 26 is shown in Figures 6.21 and 6.22.

Figure 6.21 Rain-catching system at the ventilation ridges in Hall 26. Air from the occupied space flows horizontally into the system, allowing any rain entering to be collected and drained from pans at the bottom, source: Deiter Leistner

Wind-Driven Ventilation Approaches

Wind-driven ventilation may be categorized as (1) single-sided single opening; (2) single-sided, stack-induces double opening, or (3) cross-ventilation. Single-sided ventilation describes a situation where a space is opened to the outdoors on only one side. This approach relies on air turbulence and diffusion as the primary mechanism for ventilation. Generally, this is the least effective ventilation scenario because continuity requires that the inflow volume be equal to the outflow, and in this case, both flows, which are opposite in direction, have to occur through the same opening. As is shown in the CIBSE Applications Manual AM10:1997, [16] natural ventilation in nondomestic buildings, single-sided, single-opening ventilation is applicable to a room width of less than two times the height of the space. For typical room dimensions with properly sized openings to the outdoors, this limits natural ventilation to a depth of only about 4.6 to 6.2 m (15 to 20 ft), as in Figure 6.23.

As described in the CIBSE Applications Manual AM10:1997 "single sided ventilation is driven primarily by wind turbulence." The wind-driven ventilation rate Q_w (m³/s) is given by

$$Q_w = 0.05 \, AV \,, \tag{6.1}$$

where A is the opening area (m²), and V is the wind speed at the building height (m/s).

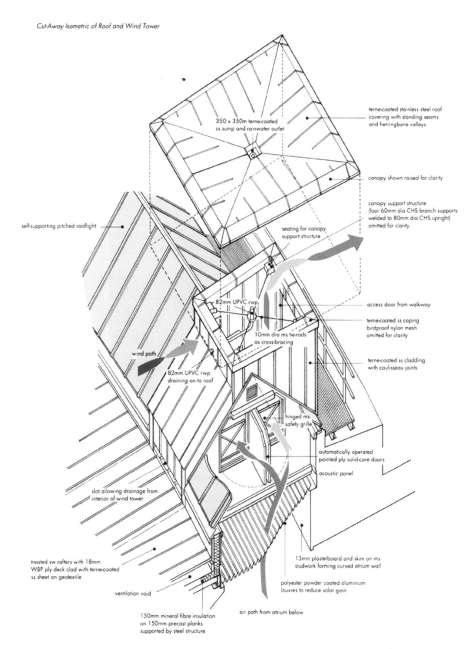

Cut-Away Isometric of Roof and Wind Tower

terne-coated stainless steel roof covering with standing seams and herringbone valleys

350 x 350m terne-coated ss sump and rainwater outlet

canopy shown raised for clarity

canopy support structure (four 60mm dia CHS branch supports welded to 80mm dia CHS upright) omitted for clarity

self-supporting pitched rooflight

seating for canopy support structure

access door from walkway

82mm UPVC rwp

terne-coated ss coping birdproof nylon mesh omitted for clarity

10mm dia ms tie-rods as cross-bracing

terne-coated ss cladding with coulisseau joints

wind path

82mm UPVC rwp draining on to roof

hinged ms safety grille

automatically operated painted ply solid-core doors

acoustic panel

slot allowing drainage from interior of wind tower

treated sw rafters with 18mm WBP ply deck clad with terne-coated ss sheet on geotextile

13mm plasterboard and skim on ms studwork forming curved atrium wall

ventilation void

polyester powder coated aluminium louvres to reduce solar gain

150mm mineral fibre insulation on 150mm precast planks supported by steel structure

air path from atrium below

Figure 6.22 Schematic of rain-catching system in Ionica building by Tim Soar, courtesy of RH Partnership

Single-Sided, Stack-Induced and Double-Opening Ventilation

Single-sided, stack-induced, double-opening ventilation can be achieved with openings on only one side of the space, but here the openings are separated vertically, as shown in Figure 6.24. The vertical separation between the openings takes advantage of thermal buoyancy (stack) effects where cool air flows in the lower opening and warm, less dense air

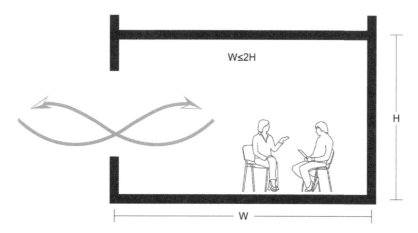

Figure 6.23 Room depth for effective ventilation with single-sided approach, source: author's figure

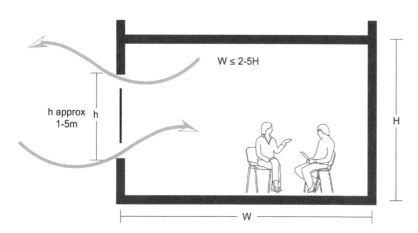

Figure 6.24 Room depth for effective ventilation with single-sided, stack-induced, double-opening approach, source: author's figure

flows out the upper opening, thus improving the overall ventilation effectiveness. The same may apply to tall single-sided openings. The CIBSE manual states that if the openings are vertically separated by about 1.5 m (5 ft) or "if the opening is tall, stack-driven ventilation can also occur." Warren and Perkins provide equations for determining ventilation rates for tall single openings or for stack-induced, double-opening systems. The stack-driven flow Q_s (m³/s) is given by [17]

$$Q_s = 0.2\,A[gh\Delta T/T_{av}]^{1/2} \tag{6.2}$$

where g is the acceleration of to gravity equal to 9.81 (m/s²), h is the height of the opening (m), ΔT is the inside-outside temperature difference (K), A is the area of the opening (m²), and T_{av} is the average of inside and outside temperatures (K). Note that degrees

Kelvin = °C + 273.16 where absolute zero in °K is −273.16 °C. For tall single-sided or single-sided, stack-induced double openings, total flow can be approximated by

$$Q_{tot} = \left(Q_w^2 + Q^2\right)^{1/2} \tag{6.3}$$

As a result, the CIBSE manual suggests that the single-sided, stack-induced, double-opening approach can extend the ventilation effectiveness to a room width of about 2.5 times the room height.

Cross-Ventilation

Cross-ventilation is the condition where inlets and outlets are located on opposite sides of the room, as in Figure 6.25. This condition is most desirable as it takes full advantage of the pressure difference between inlets on the windward side of the building and outlets on the leeward side. When inlets and outlets are near equal size, the solution is best for ventilation effectiveness, and heat or pollutant removal. The CIBSE suggests that room widths of up to five times the height can be effectively ventilated with cross-ventilation. When coupled with a central atrium with occupied spaces on both sides, building depths of over 30 m (100 ft) can be naturally ventilated.

Brown and DeKay [18] provide a graphical procedure for sizing cross-ventilation openings. As in Figure 6.26, the recommended inlet (or outlet) area as a percentage of floor area of the spacing being ventilated is a function of design wind speed and the rate of heat gain to the ventilated space from internal heat sources such as lights, equipment, and people that is to be removed by the natural ventilation, where the curves labeled 10, 20, 30, 40, and 50 represent that heat gain to the building in Btu/hr. Buildings with few occupants and relatively little gain from lights, electrical equipment, and solar gain through windows would be in the 10 to 20 Btu/ft². range, while spaces with more dense occupancy conditions and more light, equipment, and solar gain would be in the 30 to 50 Btu/ft² range. As the design wind speed increases, the required opening size decreases, and as heat gain increases, the

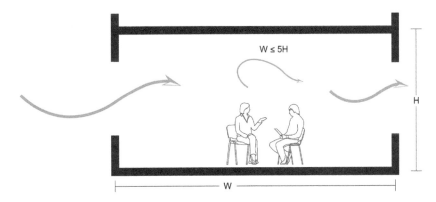

Figure 6.25 Room depth for effective ventilation with cross-ventilation. source: author's figure

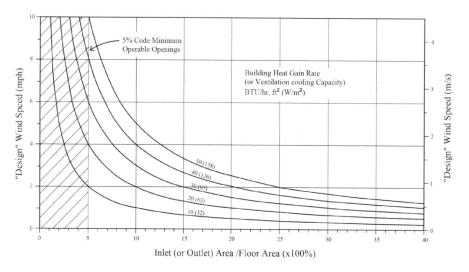

Figure 6.26 Graph for designing cross-ventilation openings, source: author's figure

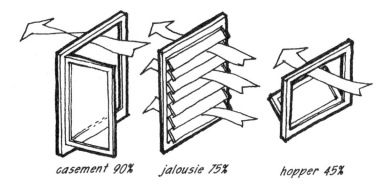

Figure 6.27 Comparison of effective opening area for alternative window types, source: author's figure

Wire Thickness Mesh Spacing	0.3 mm	0.5 mm	1.0 mm
	Geometric Degree of Obstruction		
5 mm	12%	19%	36%
8 mm	8%	12%	24%
10 mm	6%	10%	19%
20 mm	3%	5%	10%

Figure 6.28 Effective opening obstructions from wire mesh window bug screens, source: author's figure

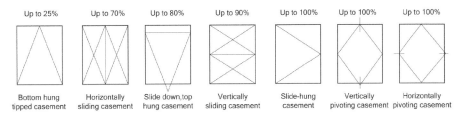

Up to 25% — Bottom hung tipped casement

Up to 70% — Horizontally sliding casement

Up to 80% — Slide down, top hung casement

Up to 90% — Vertically sliding casement

Up to 100% — Slide-hung casement

Up to 100% — Vertically pivoting casement

Up to 100% — Horizontally pivoting casement

Figure 6.29 Comparison of effective opening areas for alternative casement windows for double envelope systems, source: author's figure

opening area increases. It should be noted that 5% inlet to floor area percentage is the minimum requirement for many locations. Local wind speed conditions can be evaluated using tools such as Climate Consultant, which are available without charge from the internet.

An important aspect of determining the inlet (or outlet) opening area is to recognize that this area is the effective opening area and that windows of equal overall size may have very different effective opening areas. For example, in Figure 6.27, the casement window has the largest effective opening area at 90%, which is two times that of the hopper, slider, or single- and double-hung windows.

When window area is limited, care must be taken to select the window type that provides an adequate effective opening area. The addition of louvers or bug or bird screens can also significantly reduce the effective opening area. As shown by Oesterle et al. in Figure 6.28, [19] the percentage of obstruction to flow depends on the wire thickness and the mesh spacing, with obstruction ranging from 3% to 36%.

Oesterle et al. also suggest that when natural ventilation is coupled with a double-skin facade, the types of openings will significantly influence the ventilation effectiveness. For example, Figure 6.29 shows that various casement opening types when applied to the inner skin can result in a range of ventilation effectiveness from 25% to 100%.

Wind Towers

In addition to proper placement and sizing of windows, wind towers can become an expression of aeroform. Wind towers are design features that produce low pressure and draw air out of the building. As described in their book *Wind Towers*, Battle and McCarthy,[20] say

"The effectiveness of wind towers is dependent upon producing the maximum pressure difference between the inlet openings and the tower." They go on to say,

> The simplest design for a wind tower is a vertical construct that projects above its surroundings and has an open top. This will ensure negative pressure and provide suction in all wind directions. If the ingress of rain is a problem, a cover can be placed above the top. Alternatively, an oast wind tower (L-bend) will reduce the effect of interference at the opening and provide a greater degree of protection from the weather. However, if an oast tower is to work in all wind directions, it must be omni- directional and turn away from the wind, which obviously carries cost and maintenance implications.

They go on to say that "opening-area requirements are reduced and shorter ventilation routes may be available due to the position of the devices." This is shown in Figure 6.30.

When determining the optimum opening positions for the wind tower, Battle and McCarthy suggest,

> "The 'leading edge' of the building is the edge between the windward façade and the roof or sides. This is where the maximum positive pressure occurs – on the windward façade – and the maximum negative pressure occurs – on the roof.
>
> A wind tower is at its most effective at the windward edge of the roof (where the negative pressure is greatest), and is least successful to the leeward edge. The ideal position, which can harness winds from all directions, is therefore at the center of the building.
>
> The position of the inlets is less crucial since it is the wind tower that will drive the air through the building. The ideal position for inlet openings is along the windward façade, where the positive pressure is at its greatest. However, the change of wind direction makes this difficult to obtain without a complex building management system that can open and close windows/vents, and it is therefore better to have openings on all facades."

The wind tower may be a separate vertical element or may be integrated with a chimney or the elevator shaft and motor room. The top of the tower must be high enough

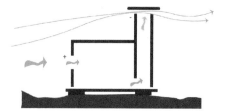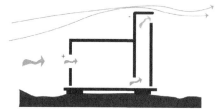

Figure 6.30 Alternative configuration of wind towers for creating outlets with low pressure, source: author's figure

to extend above the turbulence layer that develops on the roof. Also, as previously shown, a taller tower can extend higher into the boundary layer above Earth's surface where wind speeds are greater and generate lower pressures at the tower outlet. An element such as a disk (the RWE Ag building) above the tower can take advantage of the Venturi effect and further lower the pressure at the tower outlet.

The required size of the tower, in terms of total cross-sectional area, depends on a number of factors, including location, topography, amount of local wind shielding, as well as building function and internal loads. Battle and McCarthy suggest a rough guide for sizing the openings in m² is that the total cross-sectional free area of the outlets should be no less than the desired volumetric flow through the tower:

$$\text{Total air volume (m}^3\text{/s)}/1.0 \ = \ \text{m}^2\,(\text{free area}) \tag{6.4}$$

It should be noted that the minimum airflow should be at least that required for good indoor air quality as determined from ASHRAE Standard 62, typically no less than 7.0 L/s (15 cfm).

Wind Scoops

While wind towers are intended to create negative pressure and are used as airflow outlets, wind scoops catch the wind and, through positive pressurization, direct flow into the building. Wind scoops are most popular and have been used for centuries in hot, arid regions such as North Africa, the Middle East, Iran, and Pakistan. In these locations, the winds are redirected into the building while often passing over charcoal filters and a water source such as filled clay pots that evaporatively cool the air, lowering the air temperature by as much as 5°C or more before entering the building, as shown in Figure 6.31. The high mass

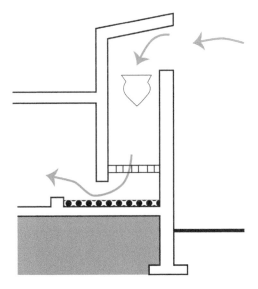

Figure 6.31 Wind tower with evaporative cooling, source: author's figure

construction of these towers also helps delay the impact of the sun and provide additional cooling (day) or warming (night) of the airstream. One design solution is the malquaf towers of Iran, Figure 6.32. These towers are compartmentalized into two or four separate vertical shafts with openings corresponding to each direction. With this configuration one face is always oriented toward the wind and experiences positive pressure while the leeward face is under negative pressure and serves as an outlet. Regional variations to the wind scoops result in different ventilation effectiveness. The Iranian four-sided malquaf performs with moderate efficiency from all directions, while the Pakistani/Egyptian variations perform best with wind directions between 0° and 60° and from about 150° to 180° incident angles, as shown in Figure 6.33.

Figure 6.32 Four-sided malquaf wind tower, source: Wikimedia

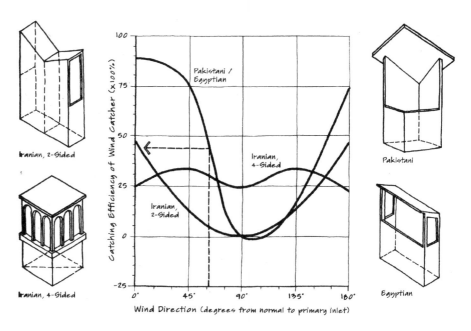

Figure 6.33 Catching efficiency for different wind catcher designs, [22] source: Khalid Abdullah Al-Megren

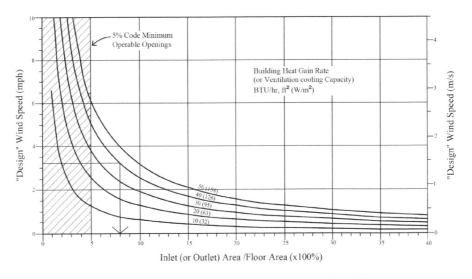

Figure 6.34 Sizing wind catchers for cooling, [23] source: author's figure

In their book *Sun, Wind and Light*, Brown and DeKay [21] show a graphical method for sizing wind catchers for cooling. Using their process, the wind catcher opening required to remove building heat gain is expressed as a percentage of floor area, with the assumption that the temperature difference between inside and out is 1.7°C (3°F), as in Figure 6.34. When using their chart, the design wind speed for a given location may be measured and averaged on-site or approximated from wind rose charts or sources, such as Climate Consultant.

Mixed-Mode Ventilation

In their book *Wind Towers*, Battle and McCarthy suggest

> that a building must not rely solely upon natural ventilation to be "green" [24], and that any use of mechanical systems compromises its integrity, is a misconception. For any number of reasons, a building may not be suitable for full natural ventilation. In these cases, natural systems can be combined with mechanical ventilation and/or air-conditioning in a mixed-mode approach that will produce considerable energy and cost savings.

They distinguish between seasonal mixed-mode and zoning mixed-mode operations. For seasonal mixed-mode operation, "if natural ventilation cannot ensure that an acceptable level of comfort is maintained throughout the year, a mixed-mode approach can be taken on a seasonal basis." With this strategy, when conditions permit, "natural ventilation is achieved through means of either wind towers, wind scoops, or a combination of both." When wind speeds are low or conditions do not support these approaches, then mechanical ventilation is used.

For zoning mixed-mode operation, "a building can be zoned to facilitate natural ventilation in certain areas while others are air-conditioned." This is often applied to buildings with atria that may not require tight thermal comfort control, and therefore, natural ventilation is applied in a zone-by-zone approach. Mixed-mode systems may incorporate the mechanical component as either an air intake (positive pressure) or exhaust (negative or suction pressure) strategy. The mixed-mode strategy is shown in Figures 6.35 and 6.36.

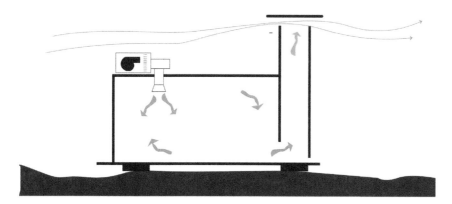

Figure 6.35 Wind tower as an outlet for mixed-mode ventilation, source: author's figure

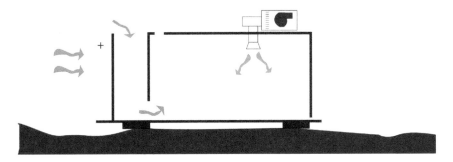

Figure 6.36 Wind tower as an inlet for mixed-mode ventilation, source: author's figure

Thermal Buoyancy (Stack)–Driven Ventilation

Pressure differences for natural ventilation may be created by wind or by thermal buoyancy. Often referred to as the stack effect or stack-driven ventilation, the ideal gas law presented in Chapter 2 suggests that as the temperature of air rises so does the volume, if the pressure remains constant. Thus, as the temperature of air rises, its volume tends to increase with a corresponding decrease in density. This decrease in air density and the effect of thermal buoyancy can be seen through the lift of a hot air balloon. As the air in the balloon is heated, it becomes less dense relative to the surrounding air mass, and the balloon lifts. The same occurs when air inside buildings is heated by internal heat sources such as lights, people, and equipment or by the sun. If inlets and outlets are properly incorporated, then airflow occurs without wind-driven or mechanical forces.

The flow of air due to stack-driven ventilation can be expressed as

$$Q = C_d A[(2/\rho_{ins})\rho_{ins}g(h \quad h_{NPL})(T_{ins}-T_{out}/T_{ins})] \, , \tag{6.5}$$

where Q is the flow rate (m³/s); C_d is the discharge coefficient (0.61 for large openings); A is the effective opening area (m²); T_{ins} and T_{out} are the inside and outside air temperatures, respectively; h is the height of the opening; and h_{NPL} is the height of the neutral pressure level (NPL) and ρins is the inside air density.

It should be noted that T_{out} is the temperature of the outdoor air entering the inlet, which may not be the same as the ambient outdoor air temperature. Equation (6.5) suggests that the effectiveness of stack-driven ventilation will primarily depend on three factors: (1) the height difference between the inlet and outlet relative to the NPL, (2) the area of the inlets and outlets, and (3) the temperature difference between the inlet (T_{out}) and outflowing (T_{ins}) air.

Generally, the greater the difference in height between the inlets and outlets, the more effective the stack-driven ventilation. This is because, as previously introduced, thermal buoyancy effects are the result of air density and therefore pressure differences in a vertical space. As the air is warmed, its density decreases, and this air tends to rise in the space. This less dense, thermally buoyant air then exerts positive pressure (outflow) near the top of the space. If allowed to escape due to the conservation of mass, this outward airflow is balanced by the inflow of cooler higher density air near the lower levels of the space. As a result of the pressure gradient from the bottom to the top of the space, high pressure develops in the upper portion of the space (pushing the air out) and low pressure in the lower portion where air flows inward due to the pressure differences with the outdoors. About midway up this space the pressure transitions from negative (inflow) to positive relative to the ambient outdoor conditions (outflow). This is shown in Figure 6.37. The height at which the pressure difference $p_{in} - p_{out}$ transitions from negative to positive relative to the ambient pressure is termed the neutral pressure level or neutral zone height. Generally, as one moves vertically, either up or down farther away from the NPL the more positive (above the NPL) or more negative (below the NPL) the pressure difference becomes relative to the surrounding pressure. Since our goal is typically to maximize the pressure difference between inlets and outlets, locating outlets as high as possible above the NPL and inlets as low as possible relative to the NPL results in maximum airflow. This is why the height difference between the inlets and outlets is an important consideration, as in Equation (6.5).

As shown in Figure 6.37, one of the consequences of stack ventilation is the NPL and the tendency for air to flow in low inlets and out of high outlets. This contributes to good indoor air quality and natural cooling. For example, if a building is designed with a naturally ventilated core shaft or central atrium with adjacent occupied spaces throughout the height of the shaft or atrium then, due to this pressure condition, cool, fresh outdoor air will flow in through spaces below the NPL, but for those spaces above the NPL, the outward flow will tend to occur where the warm, stale air from the ventilation shaft or atrium flows out through the occupied spaces. This must be considered when designing for stack-driven natural ventilation, as occupants in these upper spaces may be exposed to stale warm air. A solution is to incorporate a strategy that couples stack-driven ventilation with wind-driven pressures to create a low-pressure condition at or above the top of the ventilation shaft or

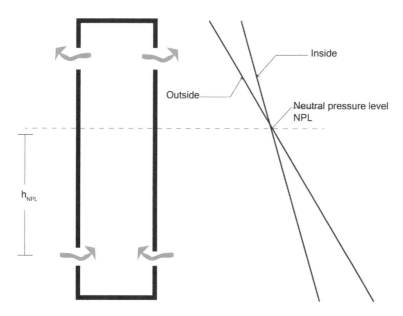

Figure 6.37 Pressure distribution due to thermal buoyancy, source: author's figure

atrium. The previously introduced inverted pyramids for the Ionica building and the airfoil disk placed above the RWE Ag building take advantage of the Venturi effect to create a low-pressure condition that shifts the NPL to the top of the building. As a result, all of the occupied spaces in the building are below the NPL, and consequently, they experience inflow; outflow only occurs at the top of the shaft or atrium. If wind cannot be relied upon to create this low-pressure condition, then mechanically assisted ventilation may be needed as a supplement.

The relative size of the inlet and outlet areas will also influence the airflow and cooling rates. Generally, if the ratio of inlet area to outlet area (or vice versa) is increased, the airflow and cooling rates will also increase, as shown in Figure 6.38. However, this increase is limited to an inlet to outlet area ratio of about 5.

Brown and DeKay provide a graphical approach to sizing openings for stack ventilation. As shown in Figure 6.39, for the same stack height, as the ratio of stack cross-sectional area to floor area increases, the heat removal rate also increases. Note that a 5% stack cross-sectional area ratio is a minimum recommended value. Also, note that increases in stack cross-sectional area to floor area is most effective for taller towers.

Equation (6.5), Figure 6.37, and Figure 6.39 show that ventilation flow and heat removal rate depend on the difference in height between the inlets and outlets. As shown in Figure 6.39, as stack height increases so does the heat removal rate. For example, for a stack cross-sectional area to floor area ratio of 12 and stack height of about 10 ft, the heat

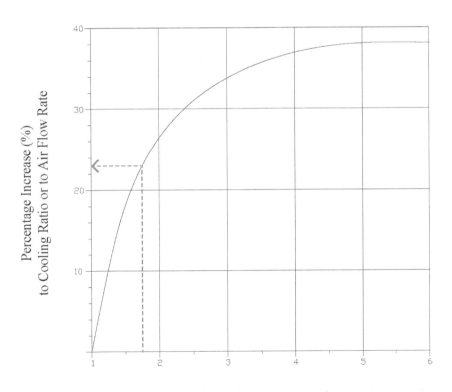

Figure 6.38 Sizing chart for inlet and outlet area ratio [25], source: author's figure

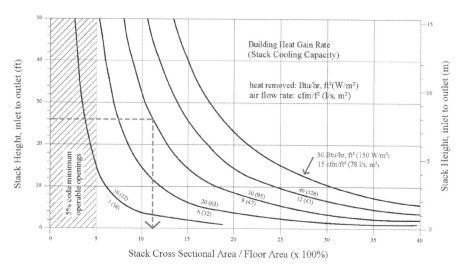

Figure 6.39 Relation between heat removal, stack height, and stack cross-sectional area to floor area, source: author's figure

removal rate would be about 20 Btu/hr, ft^2, while increasing the stack height to about 40 ft would double the heat removal rate to 40 Btu/hr, ft^2. It should be further noted that the areas are the effective area for the openings.

Equation (6.5) shows that a third determinant for the effectiveness of the stack ventilation system (Q) is the temperature difference between the inside (outlet) and outdoor (inlet) airstreams. Generally, the greater this temperature difference the greater the airflow rate will be. For inlets, this often means applying strategies to lower the temperature of the inlet air (T_{out}). This may be accomplished by evaporative cooling of the air over a body of water or water containers (such as the clay pots in the Middle eastern towers in Figure 6.31), or design elements such as a fountain, or utilization of plant transpiration to draw inlet air over grass or under a canopy of trees before introducing it into the building. In the northern hemisphere, in locations with temperate climates, this means trying to draw air from the cooler north side of a building rather than through south-facing openings.

Another strategy for cooling of inlet air (T_{out}) is ground-coupled ventilation. With this approach, before being introduced to the building interior, the outdoor air is circulated through tubes buried below grade. This approach takes advantage of the cooler, more stable temperatures of Earth when compared to outdoor air, pre-tempering the air before being introduced into the occupied areas of the building. Depending on variables, such as the depth and length of the tube, conductivity of the tubing material, and airflow rate, ground-coupled ventilation can warm the air during cold winter months and cool the air during warm summer conditions. The concept is shown conceptually in Figure 6.40, and a residential application of the strategy for Blacksburg, Virginia, is shown in Figure 6.41.

From Equation (6.5), it is desirable to increase the outflowing air temperature (T_{ins}). Therefore, to drive the thermal buoyancy for the stack ventilation outlet, the strategy is to warm the air near the top of the stack rather than cool it. Solar radiation may be employed to "boost" the outlet air temperature. For this, a radiantly transparent material such as glass or thermally conductive material can be used to warm air as it flows through the outflow

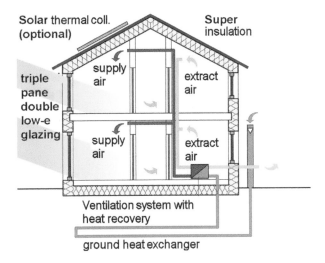

Figure 6.40 Schematic ground-coupled ventilation, source: Wikimedia

Figure 6.41 Solar CM project with ground-coupled ventilation tubes (foreground), source: author's figure

ventilation cavity. The Florida A&M University architecture school utilizes solar radiation on a ventilated sloping roof to increase the outlet air temperature and drive the ventilation scheme in the building (Figure 6.42). Similarly, the BRE environmental office building in Watford, United Kingdom, achieves cross-ventilation through the implementation of solar collecting wind towers (Figure 6.43).

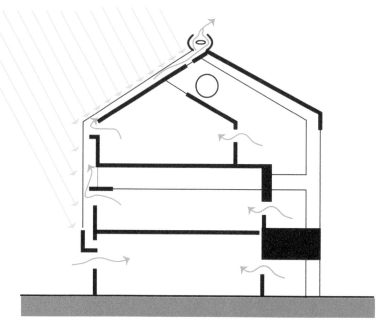

Figure 6.42 Schematic natural ventilation at Florida A&M architecture building utilizing solar heating of outlet air, source: author's figure

Figure 6.43 Ventilation shafts for BRE environmental office building, Waterford, United Kingdom, source: Wikimedia

These south-facing towers utilize glass block to transmit radiation and warm the air in the ventilation shaft. This drives the air vertically out the shaft while drawing air from the cooler north side of the building through the occupied zones.

Wind and Stack Pressures

When designing for natural ventilation, the combined effects of wind and stack ventilation should be considered. The goal of effective natural ventilation is to achieve sufficient pressure differences between inlets and outlets to induce airflow. The pressures created by wind or stack effects can be either additive or counteracting, where the goal is typically to create an additive condition. For double facade systems, this is critically important. For these systems to take full advantage of thermal buoyancy, the inlet should be low and the outlet high. However, if the external aerodynamics of the lower opening creates low wind-driven pressure relative to the upper opening, then the wind-driven pressures will counteract the thermal buoyancy pressures and flow will be reduced. This is why for projects such as the Commerzbank building, care was given to the design of the facade inlets and outlets, and aerodynamic studies were performed to ensure wind-driven pressures are high for lower openings and low for the upper openings.

Similarly, at the room or building scale, care should be taken to create conditions where wind- and stack-driven pressures are additive. For example, as previously shown, when wind flows over a sloping roof, a low-pressure condition is created. If an opening is introduced in the roof, this will help draw air out of the building. Because this opening is high, it will also serve as an outlet for stack-driven flow, and thus the two pressure-driving mechanisms will reinforce each other. On the other hand, if the roof opening is oriented into the prevailing winds, this will act as a wind scoop, and a positive pressure inflow condition will develop, as in Figure 6.44. This positive pressure will be counter to the outflow tendency from the stack-driven flow. The additive condition will result in greater flow and ventilation effectiveness.

Cross-Contamination and Compartmentalization

An important concern when designing for natural ventilation is the potential for stale air to be transferred from one occupied zone to another. With this, occupants in the downstream zones may experience discomfort from excessively warm or contaminated air that can adversely affect performance. If air is to pass from one occupied zone to another, then care should be taken to ensure the air quality is acceptable for transfer. ASHRAE Standard 62 [26] suggests that indoor zones may be categorized by "air class," as shown in Table 6.1. This classification can be useful when designing for natural ventilation, as air should not be passed from a higher to lower air class zone. Indeed, for zones classified higher than class 1, care should be taken when transferring air from one space to another.

A possible solution to this problem is to employ space compartmentalization strategies. This approach provides airflow through separate paths, usually under the

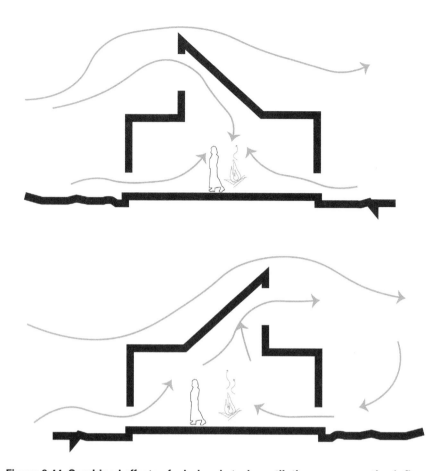

Figure 6.44 Combined effects of wind and stack ventilation, source: author's figure

Table 6.1 Room air classification

Occupancy Category	Air Class
Chemical storage/janitor closet	4
Paint spray booth	4
General manufacturing	3
Refrigerating machine room	3
Soiled laundry storage	3
General chemistry/biology lab	3
Daycare sick room	3
Kitchenettes	2
Kitchens	2
University laboratories	2
Conference/meeting rooms	1
Lobbies	1
Offices	1
Bank lobbies	1
Auditoriums	1
Classrooms	1

floor, through the structure, or above the ceiling, to prevent cross-contamination, as in Figure 6.45.

The Queens Building on the DeMontfort University campus in Leicester, United Kingdom (Figure 6.46), is a building designed for natural ventilation and compartmentalization. [27] The building functions as a teaching and research laboratory and as such must address the possible movement of contaminants from one zone to another. In both plan and section, this building is designed for airflow separation and zonal isolation. Wind towers are used throughout the project as vertical shafts for isolating airflow. Nearly every space has a dedicated inlet and outlet. The result is a project that meets the functional demands of a research laboratory while using only two-thirds of the energy of a typical laboratory.

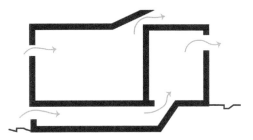

Figure 6.45 Compartmentalization strategy for ventilation flow, source: author's figure

Figure 6.46 The Queens Building, DeMontfort University Leicester, United Kingdom, source: Wikimedia

Nighttime Ventilation

Designing for airflow compartmentalization and flow through the building structure suggests an opportunity to couple natural ventilation with the building mass (high-density materials such as concrete) and take advantage of thermal inertia. Nighttime ventilation is a ventilation and energy conservation strategy that cools the mass of the building at night when outdoor air temperatures are lowest. This strategy has been historically used in hot, arid climate regions with large diurnal temperature cycles to take advantage of the cooling effects of the cool night air. By passing night air through the building and over high mass elements such as the concrete or steel structure and core of the building, the mass temperature is lowered. This pre-cooled mass acts as a thermal flywheel by absorbing early morning heat gains and thus delaying or eliminating the need for active cooling. The ventilation approach may be all natural or a mixed-mode approach. The nighttime ventilation approach removes pollutants from the indoor air while lowering energy consumption. Energy savings may range from only about 5% of the annual cooling energy to nearly 50%, depending on factors such as the climate, amount of thermal mass, building operation, and degree of convective coupling of the air with the mass. Generally, for more temperate climate zones in buildings with more thermal mass, the benefits from nighttime ventilation will increase. Benefits are usually greatest for locations with large diurnal temperature cycles.

At least two important issues for nighttime ventilation are (1) there needs to be an operational period at night when the building is unoccupied. This is because the circulation of cool night air through the building could fall below the thermal comfort zone or cause an undesirable draft. The pre-cooling of the mass may take several hours of circulation, and therefore, buildings with nighttime occupancy may have limited application for nighttime cooling. Also, care must be taken to not over cool the space so that thermal discomfort is created due to radiant exchange with the mass in the morning at the beginning of occupancy.

(2) Care must be taken to control the nighttime ventilation operation so that air with high humidity is not introduced into the building. The relative humidity of outdoor air is typically highest in the early morning hours. Circulating this moist air through the building may cause concern for condensation; effects on building furnishings and materials, such as books or paper; and possible mold growth. The operation should be controlled and limited when outdoor relative humidity conditions are high. The American Society of Heating, Refrigerating and Air- Conditioning Engineers suggests that indoor relative humidity levels should be kept below 60%, although specific operating strategies should consider issues such as building function when setting these limits. [28] In their book *Sun, Wind and Light*, Brown and DeKay give a detailed procedure for estimating the savings from nighttime ventilation. [29]

The Eastgate Building in Harare, Zimbabwe, by Pearce Partnership, with Arup engineers, takes advantage of a tropical high-altitude climate where only the two lower floors are mechanically conditioned as shown in Figure 6.47. For the natural ventilation, scheme air can be drawn through the building structure and concrete core, allowing the mass to be cooled. The majority of the thermal mass is in the ceiling, which is vaulted to increase the exposed surface area. Air enters rooms low near the window and moves diagonally back to

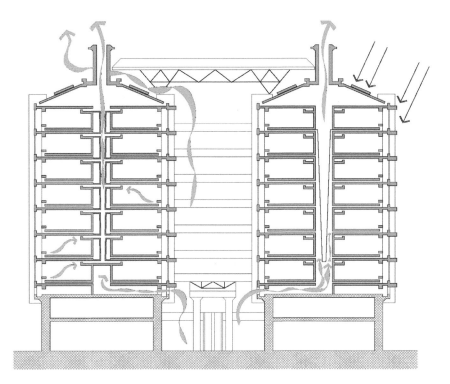

Figure 6.47 Schematic section of nighttime ventilation for the Eastgate Tower by Pearce Partnership, source: author's figure

the interior, where it is collected at high bulkheads and discharged through stack towers to outlets above the roof level. During the day, the flow is reduced through controllable louvers to a rate sufficient for fresh air supply, and the massive structure absorbs heat from internal and envelope gains. At night, airflow increases to seven air changes per hour.

As suggested by Brown and DeKay, "in nighttime ventilation schemes, the area of the mass that can be incorporated into a structure is a major limitation on the cooling potential. The ratio of mass surface area to floor area is usually between 1:1 and 1:3." Through design, this ratio should be maximized.

Final Points

A few final points for consideration should be mentioned. First, natural ventilation requires both inlet and outlet openings to the outdoors. In urban and suburban locations, these openings represent a conduit for noise penetration to the building. Strategies for reducing this noise penetration are to use inlets or outlets designed to absorb and/or reflect sound. By configuring the inlets and outlet paths to include surfaces that absorb sound, transmission to the indoors can be reduced. Similarly, surfaces such as louvers can be designed and positioned to reflect sound outwardly from the opening. This is one of the arguments for double facade systems, as they provide an opportunity to absorb and dissipate noise before it enters the building.

Two other approaches for reducing noise ingress to the building are, first, to utilize wind towers and scoops. These elements are usually tall allowing for the placement of openings to be well above the surrounding ground level.

A second strategy can be applied to medium or high-rise buildings. For these situations, the lower few floors can be closed and mechanically conditioned, while the upper floors, being well above street level, can be opened. This is a strategy for the naturally ventilated federal building in San Francisco by Morphosis where the lower four floors rely on mechanical ventilation, while the upper floors are naturally ventilated, as in Figure 6.48.

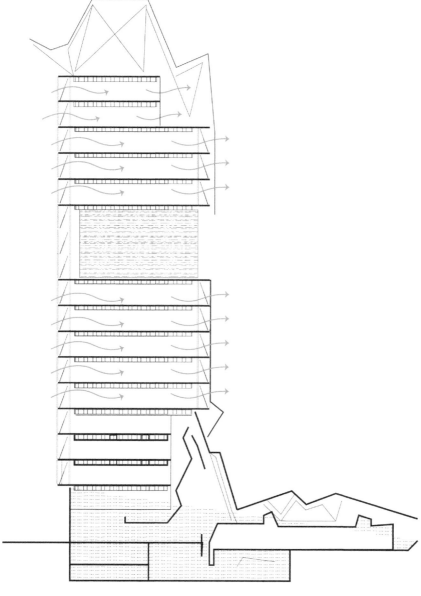

Figure 6.48 San Francisco federal building by Morphosis: section showing natural ventilation on upper floors, source: author's figure

References

[1] Zhao, Y. (2007) *A Decision-Support Framework for Design of Natural Ventilation in Non-Residential Buildings.* Electronic dissertation, College of Architecture and Urban Studies, Virginia Tech, Blacksburg, VA.

[2] Chandra, S., P. Fairly and M. Houston (1984) *Residential Cooling Strategies for Hot Humid Climates,* draft report to SERI, Cape Canaveral: Florida Solar Energy Center, as cited by Clark, G. (1990) "Passive Cooling Systems," in Cook, J. (ed.) *Passive Cooling.* Cambridge, MA, MIT Press, pp. 347–548.

[3] Bordass, W.K., K. Bromley and A.J. Leaman (1993) *User and Occupant Controls in Office Buildings: Building Design, Technology and Occupant Well-Being in Temperate Climates.* Atlanta, GA, The American Society of Heating, Refrigerating and Air-Conditioning Engineers, pp. 12–17.

[4] Chandra, S., P. Fairey and M. Houston (1984) *Residential Cooling Strategies for Hot Humid Climates,* draft report to SERI, Cape Canaveral: Florida Solar Energy Center, as cited by Clarka, G. (1990) "Passive Cooling Systems," in Cook, J. (ed.) *Passive Cooling.* Cambridge, MA, MIT Press, pp. 347–548.

[5] Heschong, L. (1999) *Thermal Delight in Architecture.* Cambridge, MA, The MIT Press.

[6] ASHRAE (1995) *Addendum to Thermal Environmental Conditions for Human Occupancy: ANSI/ASHRAE Standard 55a-1995.* Atlanta, GA, The American Society of Heating, Refrigerating and Air-Conditioning Engineers.

[7] Brager, G.S. and R. de Dear (2000) "A Standard for Natural Ventilation," *ASHRAE Journal* 42(10): 21–28. Atlanta, GA, The American Society of Heating, Refrigerating and Air-Conditioning Engineers.

[8] Oseland, N. (1998) *Acceptable Temperature Ranges in Naturally Ventilated and Air-Conditioned Offices: ASHRAE Transactions SF-98-07-4.* Atlanta, GA, The American Society of Heating, Refrigerating and Air-Conditioning Engineers.

[9] ASHRAE (2010) *ASHRAE Standard 55–2010: Thermal Environmental Conditions for Human Occupancy.* Atlanta, GA, The American Society of Heating, Refrigerating and Air-Conditioning Engineers, p. 8.

[10] Henderson, J. (1999) *Architecture for the Imagination: A Study of an Elementary Education Environment.* Master's thesis, College of Architecture and Urban Studies, Virginia Tech, Blacksburg, VA.

[11] Bowen, A. (1981) *Classification of Air Motion Systems and Patterns,* "Proceeding of the International Passive and Hybrid Cooling Conference." Boulder, CO, American Solar Energy Society, pp. 743–763.

[12] Chandra, S., P. Fairey and M. Houston (1984) *Residential Cooling Strategies for Hot Humid Climates,* draft report to SERI, Cape Canaveral: Florida Solar Energy Center, as cited by Clark, G. (1990) "Passive Cooling Systems," in Cook, J. (ed.) *Passive Cooling.* Cambridge, MA, MIT Press, pp. 347–548.

[13] Jones, D.L. (1998) *Architecture and the Environment: Bioclimatic Building Design.* Woodstock, NY, The Overlook Press, pp. 170–173.

[14] Baker, N. and K. Steemers (2000) *Energy and Environment in Architecture: A Technical Design Guide.* New York, NY, E&FN Spon Publishing, pp. 170–179.

[15] Daniels, K. (1997) *The Technology of Ecological Buildings.* Boston, MA, Birkhauser Verlag, pp. 110–112.

[16] CIBSE (1997) *Natural Ventilation in Non-Domestic Buildings – Applications Manual AM10: 1997.* London, UK, The Chartered Institution of Building Service Engineers, Bath Midway Press.

[17] Warren, P.R. and L.M. Parkins (1985) *Single-Sided Ventilation Through Open Windows: ASHRAE SP491.* Available from: www.ornl.gov/sci/buildings/2012/1985B3papers/015.pdf.

[18] Brown, G.Z. and M. DeKay (2001) *Sun, Wind and Light: Architectural Design Strategies,* 2nd edition. New York, NY, John Wiley and Sons Inc., pp. 182–190.

[19] Oesterle, E., R.D. Lieb, M. Lutz and W. Heusler (2001) *Double-Skin Facades.* Munich, Germany Prestel Verlag, p. 102.

[20] Battle McCarthy (1999) *Wind Towers.* New York, NY, John Wiley & Sons.

[21] Brown, G.Z. and M. DeKay (2001) *Sun, Wind and Light: Architectural Design Strategies,* 2nd edition. New York, NY, John Wiley and Sons Inc., p. 187.

[22] Al-Megren, K.A. (1987) *Wind Towers for Passive Ventilation Cooling in Hot-Arid Regions.* Arch D. thesis, University of Michigan, University Microfilms International, Ann Arbor.

[23] Brown, G.Z. and M. DeKay (2001) *Sun, Wind and Light: Architectural Design Strategies,* 2nd edition. New York, NY, John Wiley and Sons Inc., p. 190.

[24] Battle McCarthy (1999) *Wind Towers,* New York, NY, John Wiley & Sons, p. 40.

[25] Brown, G.Z. and M. DeKay (2001) *Sun, Wind and Light: Architectural Design Strategies,* 2nd edition. New York, NY, John Wiley and Sons Inc., p. 187.

[26] ASHRAE (2010) *ASHRAE Standard 62.1–2010: Ventilation for Acceptable Indoor Air Quality.* Atlanta, GA, The American Society of Heating, Refrigerating and Air-Conditioning Engineers.

[27] Battle McCarthy (1999) *Wind Towers.* New York, NY, John Wiley & Sons, pp. 62–65.

[28] ASHRAE (2010) *ASHRAE Standard 55–2010: Thermal Environmental Conditions for Human Occupancy.* Atlanta, GA, The American Society of Heating, Refrigerating and Air-Conditioning Engineers, p. 7.

[29] Brown, G.Z. and M. DeKay (2001). *Sun, Wind and Light: Architectural Design Strategies,* 2nd edition. New York, NY, John Wiley and Sons Inc., p. 193.

Chapter 7

Forced Airflow

Airflow may be categorized as either mechanical, wind, or buoyancy driven. When resource conservation is a design goal, as it often is, wind- or buoyancy-driven natural ventilation is often desirable and has great advantages, but in many situations, ventilation by natural means cannot provide the desired levels of comfort. Mechanical ventilation can provide adequate control of temperature at comfortable levels, and ample exchange of inner space air with outdoor air. The American Society of Heating, Refrigerating and Air-Conditioning Engineers (ASHRAE) defines mechanical ventilation as "*ventilation provided by mechanically powered equipment, such as motor-driven fans and blowers (Figure 7.1), but not by devices such as wind-driven turbine ventilators or mechanically operated windows.*" [1]

Examples of mechanically driven airflow in buildings include air movement from space to space for the purpose of HVAC (Figure 7.2); supply of outdoor air for ventilation (Figure 7.3); exhaust airflow for contaminated air removal from fumehoods, laboratories, restrooms, kitchens, and janitor closets (Figure 7.3); heat removal by roof and attic ventilation (Figure 7.4); and envelop ventilation for heat removal (Figure 7.5). With such common applications of mechanically driven airflow, knowledge of this should be part of an architect's and architectural engineer's education.

Forced airflow can have a number of formative influences on architectural design. Louis Kahn is arguably one of the most influential American architects with notable projects including the Exeter Library, the Salk Institute, and the Richards Medical Research Laboratory. Perhaps his most famous design is the Kimbell Art Museum located in Fort Worth, Texas. The Kimbell has repeatedly been recognized as one of the best art museums in the world. In the Kimbell, the play of light and shadow gives the interior depth with lighting that is appropriate for a gallery. But the consideration of forced air movement also contributes

 DOI: 10.4324/9781003167761-7

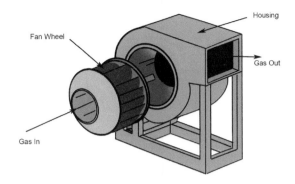

Figure 7.1 Typical centrifugal fan components, source: author's figure

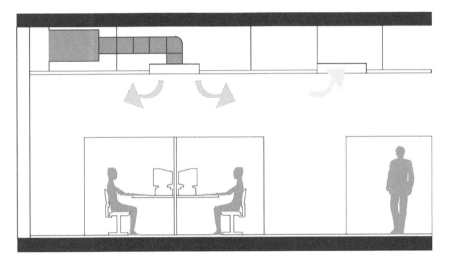

Figure 7.2 Room air distribution, source: author's figure

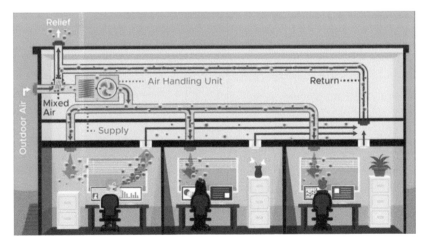

Figure 7.3 Schematic outdoor air mechanical ventilation distribution, source: EPA

Figure 7.4 Attic ventilation fan (courtesy of SolaTube)

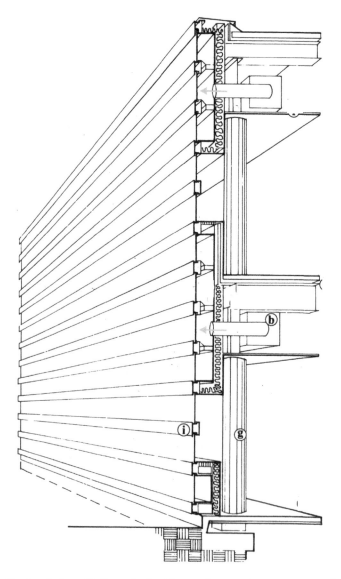

Figure 7.5 Mechanical ventilation of a double envelope system, source: author's figure

Figure 7.6 Kimbell Art Museum by Louis Kahn, source: Wikimedia

to the building form. Kahn is well-known for his architectural proposition for served and servant spaces. Served spaces are inhabitable, and where the programmatic functions take place. The servant spaces are dedicated to the building services: plumbing, lighting, and air-flow. In the Kimbell, the distinction between served and servant space becomes formative as the structural separation between the main roof vaults is also used to distribute forced air throughout the building (Figure 7.6 and Figure 7.7). This chapter is dedicated to understanding fundamental issues of forced air systems.

Inlets and Exits

Air-conditioned air is usually directed through pipes and ducts. This implies that at some point, air will enter a machine or just an opening of a duct, and at another point, air will be discharged in an interior space or outdoors. There are some basic features of the motion of air going into intakes and coming out of exits. These hold for any fluid, which includes water and air. These flow features have been proved mathematically and verified by multiple experiments. The most basic features of flow into intakes and out of exits are the following:

a. Flow from an open space into an intake is collected from the entire space. In other words, every fluid element in the space is set in motion toward the intake.
b. Flow emerging from an exit into an open space is directed along straight lines and contained in a stream known in fluid mechanics as a **jet**.

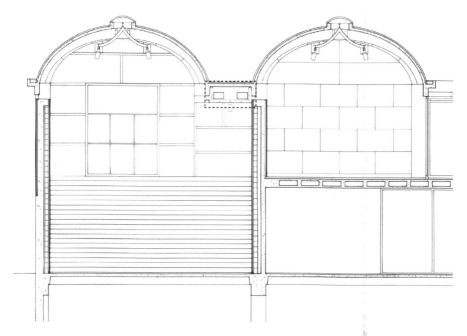

Figure 7.7 Kimbell section showing HVAC ducts, source: author's figure

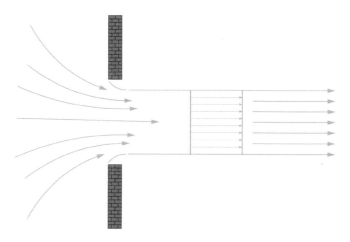

**Figure 7.8 Flow through a hole on a wall containing a fluid under pressure, source:
author's figure**

These flow properties are represented schematically in Figure 7.8 where we show a wall
containing on the left a fluid and on the right an open space. The space on the left could be a
big tank containing a liquid and on the right open air. The liquid would then be collected from
all points in the container and will flow straight as a jet into the open space. The space on
the left could also be a tank containing a gas under pressure. Again, the gas will be collected
from every point in the tank (or room) and will flow as a jet to the right.

Another feature of the flow through a hole in the wall is something we have seen before. Fluids cannot make sharp turns. Streamlines form a smooth curve out of the hole and form a jet with a diameter smaller than the diameter of the hole. Also, at least close to the exit, the velocity profile is uniform. In case a fluid is sucked into an inlet on a flat wall (Figure 7.9), again, streamlines separate at the inlet corners but their expansion to the size of the duct further downstream introduces turbulence.

It should be noted and emphasized here that turbulence in ducts leads to losses of the efficiency of operation, as well as generation of noise. Most vendors for air distribution components such as room air grilles or diffusers provide the noise rating (NR) for their products, which indicates the increase in sound pressure level as the result of airflow through the device. In rooms such as classrooms or performance spaces such as auditoriums, where background noise levels should be kept low for good communication, the comparative NR can be important selection criteria when considering alternative air diffusion products. To reduce the level of intake turbulence and NR, intakes can be formed to guide the flow through smooth turns. A good example is the bell mouth intake shown in Figure 7.10. If the space is not available for such a design, or its cost is prohibitive, intake turbulence could be reduced by a set of straightening vanes as shown in Figure 7.11. These vanes are indicated with a set of thick orange straight segments. In practice, the effect of straightening vanes can be achieved with honeycombs, which are commercially available.

When fluid is discharged from a hole on a tank, as shown in Figure 7.12, or from a pipe, the stream initially does not expand. The flow moves more or less along straight lines normal to the wall of the container or along the axis of the discharging pipe. Actually, some mixing with the ambient fluid does occur downstream, as shown schematically in Figure 7.12. Fluid moving in a long pipe may have developed turbulence and a nonuniform profile, but both characteristics have little effect on the direction of the jet. However, the sheer character of the flow right next to the wall could induce the development of large turbulent structures indicated in the figure in the form of rolling streamlines. These induce some mixing with the ambient fluid, indicated in Figure 7.12 with a velocity profile, which expands beyond the extension of the pipe walls.

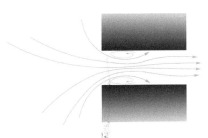

Figure 7.9 Duct inlet on a flat wall, source: author's figure

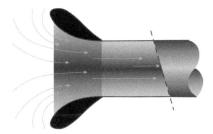

Figure 7.10 Bell mouth inlet, source: author's figure

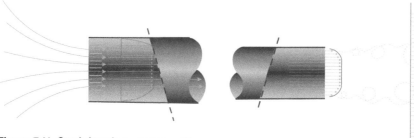

Figure 7.11 Straightening vanes allow a smooth turning of the flow into an intake, source: author's figure

Figure 7.12 Stream issuing from a pipe with velocity profile, source: author's figure

A jet flowing into an open space extends far from the exit, but eventually, it dissipates. This means that its kinetic energy is converted to heat, at levels that cannot be detected. But in small inner spaces, like a small room, a jet may reach the opposite wall, and after hitting the wall, it may turn along the wall and continue in the form of a wall jet. This may be desirable with HVAC systems such as displacement ventilation that introduce the air low in the space and extract the air high on the opposite side of the room. Because the air motion is first out from the diffuser then up, displacement systems are efficient at moving contaminants up and out of the breathing zone and are sometimes used in hospital patient rooms. However, care must be taken to not introduce draft to the feet of the room occupants.

Jet nozzle diffusers are frequently used in buildings such as rail stations and airport terminals. Here, since the height of these spaces is typically high, jets of air are discharged above the occupied zone so as to not blow directly on occupants and create draft. The air jet can thus extend far into the space to provide good air distribution. When cooling, this air jet naturally mixes with the entrained room air and, because the temperature of the air is relatively low, the air falls into the occupied zone at low velocity, thus maintaining comfort conditions. These nozzle diffusers are often visible throughout the terminal as in Figure 7.13.

Flow in a Closed Space

Body heat is released from human bodies by natural or forced heat convection. The body heats the air in its immediate environment. Air slightly warmer than the ambient air tends to rise as a plume off the top of one's head, and thus it creates a small stream that further removes heat. Forced heat convection happens when a stream of air moves over a human body. If this stream is cool air, then cooling of the body is more effective but could be unpleasant, as with the case of draft. This approach is common for the personal jet nozzles on airplanes. On the other hand, if the stream is hot, heating a room in a cold environment, then it would have an opposite effect. In other words, the

Figure 7.13 Use of nozzle jet diffuser, source: Holyoake

Figure 7.14 Exit and return in a closed space, source: author's figure

forced stream will remove the heated air generated by the body faster than the heat it can offer to the body. The final conclusion we can draw from this discussion is that a designer should avoid air-conditioning jets aimed directly at areas where occupants of a space may be sitting.

There is another negative issue associated with concentrated jets in a closed space. All air-conditioning systems circulate the air through heat exchangers and filters. So, air discharged in a space must be returned or exhausted, and this is achieved by installing an outlet in the same space. A concentrated discharge jet could be trapped in a short-circuited path and leave the room without exchanging the heat as desired. This could happen if the return is on a wall opposite the jet exit or on the same wall but very close to the exit, as shown schematically in Figure 7.14. This might also occur if ceiling-mounted supply diffusers and return air grilles are placed too closely together. The inlet and exit designs of this figure should be avoided. This is also a concern for

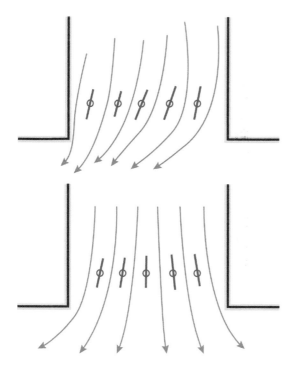

Figure 7.15 Ceiling duct vanes, source: author's figure

ventilation as presented in ASHRAE Standard 62: Ventilation for Acceptable Indoor Air Quality. Nearly all nonresidential buildings require that outdoor air be supplied to the indoors, thus typically some or all of the air being supplied to a room is outdoor air passing through the HVAC system. For maintaining good indoor air quality, it is important that this ventilation air get to the breathing zone and be well-mixed in the room and not short-circuited, as described earlier.

Commercial systems are available to disperse and spread a jet in a closed space. The most common system consists of a set of fixed vanes. The inner vanes are adjustable and can direct the flow in one direction. The outer set of vanes is attached to the ceiling and can just spread the flow in two directions, as shown in Figure 7.15. There is very little the user can do with this system, and the aim should be to adjust the inner set of vanes away from the space where occupants may be sitting.

Another system is available in larger spaces, mounted on a wall, and most often found in hotels. These involve adjustable vanes that can be set at different angles individually (Figure 7.16). Such vanes can be used to spread the jet horizontally, while another set of adjustable vanes can direct the jet to the ceiling to avoid hitting occupants of the room. Amazingly, the authors have observed that these vanes are very seldom adjusted in public spaces or in hotels.

In architectural applications, the previous scenarios fall mainly into two device categories – namely, grilles and diffusers. The most efficient device to spread a jet issuing from an exit is a diffuser (Figure 7.17). ASHRAE describes

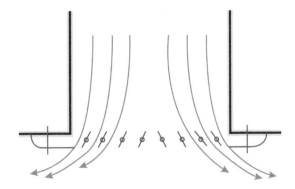

Figure 7.16 Individually adjustable vanes, source: author's figure

a supply air grille as consisting of a frame enclosing a set of either vertical or horizontal vanes (for a single deflection grille) or both (for double-deflection grille). These are typically used in sidewall, ceiling, sill, and floor applications. A diffuser, on the other hand, usually generates a radial or directional discharge pattern. For ceiling applications, this pattern is typically parallel to the mounting surface. Diffusers may also include adjustable deflectors that allow discharge to be directed perpendicular to the mounting surface. A diffuser typically consists of an outer shell, which contains a duct collar, and internal deflector(s), which define the diffuser's performance, including discharging pattern and direction. [2]

Diffusers comprise a set of fixed vanes that are arranged in four groups, each directing the flow outward. A jet is thus broken into four jets, each pointing away from the direction of the duct that delivers the stream. An alternative design is one with circular vanes that spreads the flow in all directions. Diffusers can be specified to throw the air in four, three, two, or only one direction depending on the location within the room. Diffusers may also be linear to be integrated with common suspended ceiling tiles.

Diffusers are often located on the ceiling and discharge the air horizontally, nearly parallel with the plane of the ceiling.

An important design consideration for the grille or diffuser selection is the **throw**. Diffusers are typically specified by two characteristics, discharge velocity and throw distance. The *throw is the distance from a center of the outlet face to a point where the velocity of the air stream is reduced to a specified velocity, usually 150 [0.75], 100 [0.50] or 50 fpm [0.25 m/s].* [3] The specified velocity is referred to as the **terminal velocity**. Generally, the throw will be larger for higher discharge velocities, and the velocity will be less the further from the diffuser. Understanding the throw is important for such design decisions as how far apart to place the grilles or diffusers to maintain good air distribution, where generally the throw should be approximately half of the spacing distance between devices.

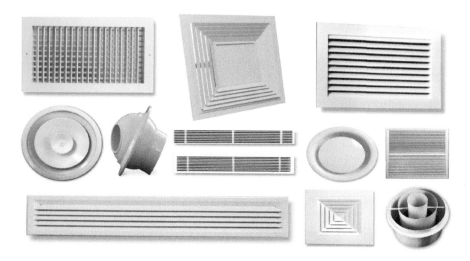

Figure 7.17 Ceiling diffusers, source: Clare White

This airflow pattern often results in a Coanda effect. The Coanda effect is the phenomenon in which a jet flow attaches itself to a nearby surface and remains attached even when the surface curves away from the initial jet direction. The phenomenon derives its name from the aeronautical engineer – Henri Coanda. The Coanda effect can increase the throw distance of a jet of air from the diffuser and thus allow for greater spacing distances between diffusers.

The selection and specification of grilles and diffusers depend on a number of architectural considerations such as the following:

a.   How many grilles or diffusers will be required for a given room geometry and the *throw* of the devices used?
b.   What will be the relative location of grilles and diffusers, and the return outlets for good air distribution?
c.   Do the grilles or diffusers need to be adjustable?
d.   Is draft a concern for air discharge directly on the occupants?

Flow in Ducts and Pipes

Ducts and pipes carry fluids from a source or a container and deliver them to where they are needed. The most common systems are the piping systems that deliver water to kitchens and bathrooms, and ducts that deliver conditioned air for heating, cooling, and ventilating indoor spaces.

The motion in any pipe or duct system can be analyzed mathematically by the energy equation. This equation is based on the principle of energy conservation and was derived analytically. It states that the energy entering a system at a station 1, is equal to the losses or gains that the flow experiences through the pipe plus the energy that leaves the system at the station 2. Expressed in terms of energy per unit fluid volume, the equation reads, White [4]

$$p_1 + \rho V_1^2/2 + \gamma z_1 + \gamma h_p = p_2 + \rho V_2^2/2 + \gamma z_2 + \gamma h_f + \gamma \Sigma h_m \qquad (7.1)$$

This is essentially Bernoulli's equation, which we have seen in Chapter 3, as Equation (3.5), to which we have added terms to account for losses and the effect of a pump or a compressor. In this equation, the new terms represent the following:

(γh_p): energy offered to the system by the pump. Here h_p is the head of the pump or blower.

(γh_f): energy lost as the fluid moves through the pipe. This is essentially fluid friction.

($\gamma \Sigma h_m$): losses due to special system components. The summation sign indicates the sum of minor energy losses.

We will discuss these terms in some detail and provide tables or graphs for their numerical calculation. A figure that schematically displays some of the terms in this equation is presented in Figure 7.18. The elevations are shown measured from a reference level, which can be arbitrarily defined. The choice of the reference level has no effect on the final answer since it is the difference between z_1 and z_2 that enters in the calculation.

We get a more practical version of this equation if we divide by the specific weight, factor γ.

$$p_1/\gamma + \rho V_1^2/2\gamma + z_1 + h_p = p_2/\gamma + \rho V_2^2/2\gamma + z_2 + h_f + \Sigma h_m \qquad (7.2)$$

The physical meaning of each term in this equation has not changed. The factor γ is just a constant, but this constant has dimensions, and thus only the dimensions of terms in this equation have changed. But in this form, we can recognize quantities that are easier to grasp in practice. Each term in the equation now has units of length. And some of these lengths can be readily identified, such as the elevation at station 1, z_1, and station 2, z_2; and the head of the pump, h_p.

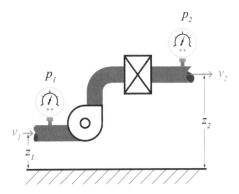

Figure 7.18 Schematic drawing indicating terms of the energy equation, source: author's figure

Some of the energy losses are due to friction. The shear stresses that develop along the duct or pipe wall tend to obstruct the fluid motion in a duct or a pipe. These losses are proportional to the length of the pipe or the duct, L; inversely proportional to the pipe diameter or duct width, d; and can be expressed as follows:

$$h_f = (fL/d)V_2^2/2 . \tag{7.3}$$

Here f is the friction coefficient, which in turn depends on the flow velocity, V; the fluid density, ρ; the kinematic viscosity, ν; and the roughness of the walls. The friction coefficient can be found in graphs and tables given in most fluid mechanics books. Actually, for fully turbulent flow, f can be approximated by the following equation:

$$f = \left[2\log(\varepsilon/3.7d)\right]^2 , \tag{7.4}$$

where ε is the relative roughness of the walls, a quantity given by the manufacturer of the pipe or duct. For the flow of air in air-conditioning ducts, the wall roughness plays a negligible role, and the friction coefficient can be approximated by the numerical value of 0.01.

Energy is lost as the flow is moving through valves, bends, and other pipe and duct components. This is because the flow separates, as described, for example, in Figure 7.8, and forms a pocket of recirculating flow, which breaks and forms a number of large and small vortices. The kinetic energy contained in these vortices cannot be recovered, and the corresponding losses known as minor losses must be included in the energy equation. A large number of experiments have been conducted for various configurations to provide estimates of the minor losses, and the results are expressed in terms of constant coefficients denoted by the symbol k.

$$h_m = kV^2/2 \tag{7.5}$$

Specific values of the minor loss coefficients for different types of fittings are included in the section on air-conditioning ducts that follows.

Of interest to engineers and architects in systems represented by the energy equation is the volume flow rate, Q, which is related to the flow velocity by the product VA, where A is the cross-sectional area of the pipe.

$$Q = VA \tag{7.6}$$

The head of a pump is proportional to the power needed to move the fluid. The power is also proportional to the flow rate Q.

$$P = \gamma Q h_p \tag{7.7}$$

It should be noted here that the power expressed by this equation is the power required to move the fluid. But the power required by the motor or engine driving the pump, known as the brake-horse power, bhp, is larger because of mechanical losses in the machinery.

This can be expressed by an efficiency coefficient, η, so that the bhp, the power required to drive the pump, is related to P by the formula

$$bhp = P/\eta .$$
(7.8)

The designer can choose Points 1 and 2 in Equation (7.2) in any way that would make the calculation easier. This equation can be solved for one unknown quantity. If for example one needs to find the head of the pump, then one should enter the initial and final pressure, velocity, and elevation, as well as the losses.

A very simple case that can demonstrate the use of this equation is the calculation of the head of a pump to move water from a container to another tank at a higher elevation, as shown in Figure 7.19.

Here we can choose our Points 1 and 2 to be right at the free surfaces. The advantage of this choice is that if these containers are large, the velocity at the free surface is practically zero, $V_1 = V_2 = 0$. Moreover, since the free surfaces are exposed to the atmosphere, p_1 and p_2 are equal to the atmospheric pressure and thus equal to each other, $p_1 = p_2 = p_0$, and can be canceled. We can further simplify the equation assuming that the losses are negligible. Equation (7.2) then becomes

$$z_1 + h_p = z_2 .$$
(7.9)

Now the action of a pump becomes clear, since its head is equal to the physical head of z_2-z_1. The pump head is thus directly related to the volume flow rate via Equation (7.10).

$$h_p = z_2 - z_1 = P/(\gamma Q)$$
(7.10)

So, for a given head z_2-z_1 and a required flow rate Q, we can find the power of the pump needed,

$$P = \gamma Q(z_2 - z_1) .$$
(7.11)

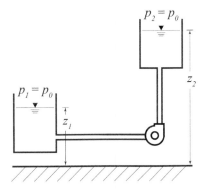

Figure 7.19 A simple application of the energy equation, source: author's figure

If we are dealing with air, as in the case of forced or natural ventilation and air-conditioning, the energy equation is somewhat simplified. For elevations the order of even a few hundred feet, the terms γz_1 and γz_2 are much smaller than all other terms and therefore can be dropped from Equation (7.1). The energy equation then becomes

$$p_1 + \rho V_1^2/2 + \gamma h_p = p_2 + \rho V_2^2/2 + \gamma h_f + \gamma \Sigma h_m . \tag{7.12}$$

The consideration of these pressure losses in architectural design can be seen in high-rise buildings. For tall buildings, intermediate floors that house the air handling units can often be seen every ten floors or so in height. These mechanical floors often appear different from the occupied floors because outdoor air for ventilation and exhaust air must be exchanged with the outdoor, resulting in a more porous enclosure system. By designing with mechanical floors every ten floors or so in height, long duct runs are avoided, and the horse power of the motors moving the air can be smaller.

Pressure losses in ducts are also a concern when laying out the floor plan for large buildings where multiple mechanical rooms should be distributed to serve different HVAC zones. This is very common for buildings such as schools where different HVAC systems would serve the administrative offices, the classrooms, the gymnasium, and the cafeteria. The inclusion of HVAC rooms should be part of the early design development stages.

Fans and Blowers

Mechanically driven airflow, as the term implies, is driven by a pressure difference produced by a motor-driven device such as a fan or blower. Fans and blowers may be categorized as axial (Figure 7.20) or centrifugal (Figure 7.21), where centrifugal fans may be subcategorized as forward curved, backward curved, and radial (Figure 7.22). Centrifugal fans are usually referred in practice as blowers.

Figure 7.20 Axial fan, source: Wikimedia

Figure 7.21 Centrifugal fan, source: Wikimedia

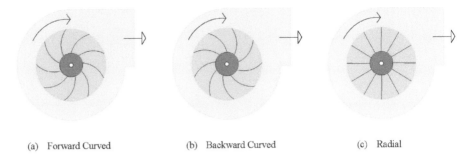

(a) Forward Curved   (b) Backward Curved   (c) Radial

Figure 7.22 Types of centrifugal fans, source: author's figure

In mechanically driven systems, the fan or blower is essentially a pressure-driving device that creates a pressure difference between the inlet, low pressure, and outlet, high pressure. As a result, this creates pressure differences throughout the building, where the supply air from the fan is at a higher pressure than the return air flowing back to the system. In smaller applications, such as residential and small commercial spaces, one fan provides the pressure difference. In larger buildings, both a supply and a return fan might be used together. This is because the horsepower and electrical consumption are greater for one large fan pushing and pulling the air, as opposed to one fan (supply) pushing the air with the other, return, pulling the air.

Residential systems typically supply and return a constant flow of air from the space through the system. For residential systems, air is typically supplied at the perimeter of the house, often near windows, and returned near the center of the space in a central hallway. Because the return air is well-mixed, the thermostat is often located near the return air intake so as not to be influenced by localized heat losses or gains, or direct sunlight from the perimeter.

Fans and blowers are characterized by the head, h_p, and by the volume flow rate, Q, they produce. The head is the pressure difference between the intake of the machine and its discharge side.

The head is the quantity h_p that appears in Equation (7.2), and Q is the quantity defined in Equation (7.6). These features are not fixed for a certain machine. They depend on each other, so that in most such machines, if the load – i.e., the head – required in a certain application is larger, then the volume flow they deliver is smaller.

Manufacturers provide the characteristics of their machines together with the corresponding specifications. Most important of this type of information are the head-flow rate graphs. In Figures 7.23 and 7.24, we present typical such graphs that show the general character of blowers and fans. Actual machine graphs display numerical values of head and flow rate on the axes of a graph.

The most dominant feature is that as the head increases, the flow rate delivered decreases. Blowers are actually a little more forgiving than axial fans, because for a good range of the flow rate there is little change in the head. In fact, for a range of flow, the head increases with the flow rate, and this trend is more pronounced in forward-curved impeller blades. On the other hand, the ability of blowers to withstand a load drops precipitously at their higher range of flow. But axial fans have a smooth drop of head with the increase of the flow.

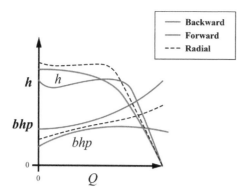

Figure 7.23 Head-flow rate graph typical of blowers, source: author's figure

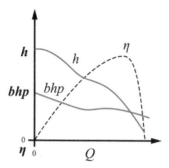

Figure 7.24 Head-flow rate graph typical of axial fans, source: author's figure

Among the other useful information concerning fans and blowers are the ideal fan laws. These laws relate changes in flow to the rotational speed and horsepower of the motor. Two of the ideal fan laws are presented in Equations (7.13) and (7.14). In Equation (7.13), the change in flow rate (CFM) is related to the rotational speed of the motor (RPM). This law states that the ratio of CFM is proportional to the ratio of speeds. For example, to double the flow one needs to double the machine speed. In Equation (7.14) the rotational speed of the motor is related to the horsepower. But now, the ratio of the power is proportional to the cube of the ratio of the speeds.

$$CFM_2 / CFM_1 = (RPM_2)/(RPM_1) \qquad (7.13)$$

$$HP_2 / HP_1 = [(RPM_2/(RPM_1)]^3 \qquad (7.14)$$

What is very interesting is that these equations can be rearranged, as in Equation (7.15). Here, the change in horsepower is shown to not be linearly related to the change in flow (CFM) but is related to the cubic power of the change in flow. In other words, to reduce the airflow in a room to ½ of the initial flow rate would require only $(1/2)^3 = 1/8$ the initial horsepower. This cubic relationship of flow to horsepower is the key to why variable air volume

(VAV) HVAC systems have become the most common system application. For a VAV system, as the cooling loads in a building change throughout the day due to changes in heat gain, the volumetric flow also changes. As shown in Equation (7.15), the cubic relationship of change in flow to horsepower can result in significant energy savings when the system is operating at low flow, as is often the case.

$$HP_1/HP_2 = \left[CFM_1/CFM_2 \right]^3. \tag{7.15}$$

For most nonresidential buildings in the United States, in addition to heating and cooling, the air handling system must provide ventilation (fresh outdoor) air throughout the occupied zones. To balance this intake of outdoor air through the system, air is relieved (exhausted) from the building, usually at the HVAC system. When the ventilation airflow and relief flow are balanced, then the pressure difference between the inside and out gets close to zero. However, in many situations, the HVAC system is set up to intentionally bring in slightly more outdoor ventilation air than relieved or exhausted. This puts the building under slightly positive pressure when compared to the outdoors, which reduces or eliminates infiltration. Under this situation, the only air entering the building is at the outdoor air inlet of the HVAC system where the air can be heated or cooled as needed, as well as filtered. Improperly balanced building ventilation can often be recognized when the entry doors are difficult to open, which indicates that the building is under negative pressure. This would result in uncontrolled infiltration.

The distribution of air throughout the building as well as considerations for where to bring in the outdoor ventilation air or exhaust the stale air can be formative. For example, the Richards Medical Research Laboratory building by Louis Kahn is an example of the formative influence of forced air movement. Because the facility is a research lab, the locations of ventilation and exhaust airflow were major considerations. For this, Kahn introduced vertical shafts to accommodate the air distribution, as shown in Figure 7.25. The vertical shafts became dominant features for the building's massing and elevation. In addition, the air shafts, along with the stairwells, provided balance and symmetry in the floor plan and massing.

In buildings with a high level of concern for indoor air quality and contamination containment, an exhaust fan may be added to the system. The exhaust fan moves air from indoors to out, and commonly exhausts air from spaces such as kitchens, restrooms, and janitor closest. In hospitals and laboratory buildings, where containment of infectious diseases or spillage of harmful agents need to be contained, a fumehood and/or exhaust fan can be used to create a pressure situation in which the contaminated space is at low pressure relative to the surrounding spaces, and therefore air flows into the containment space. This can be achieved by an independent AC system that sucks air from the contaminated space, thus reducing the pressure in it and discharging it safely through filters outdoors.

When indoor space must be protected from any possible contamination or particulate infiltrating through cracks and minute openings, the opposite air control system is needed. The internal pressure must be kept higher than all neighboring rooms. This is called a clean room, or a **cleanroom**. A special mechanical system that includes pumps and multiple filters is required. When indoor air quality is very important, the airflow paths can be manipulated by active air quality and pressure monitoring, as in Figure 7.26.

Figure 7.25 Richards Medical Research Laboratory with dominant airflow shafts, source: Wikimedia

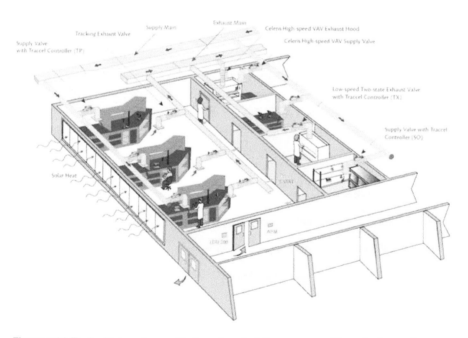

Figure 7.26 Typical laboratory with fumehood airflow, source: Phoenix Controls

One important architectural design consideration for exhaust systems is to provide a path, both horizontally and vertically, for the air to move from the occupied space to the outdoors. For example, restrooms and janitor closets, which are exhausted spaces, must have a path for the air to move to the outdoors. This often means that the floor plan must include "**chases**," which are vertical shafts in the building for plumbing and HVAC distribution. While these chases may not be visible or accessible by building occupants, they do require careful planning from the early stages of design.

In nearly all HVAC systems, the air is moved to and from the occupied spaces through conduits referred to as ducts. While the duct provides a path for air movement, as previously introduced, it also introduces resistance to the airflow. Generally, the greater the volume of flow in relation to the surface area of the duct, the lower will be the resistance. It should be recalled that flow friction in a conduit is proportional to its surrounding area. For the same contained area, a circle has the smallest periphery of any other shape. And thus, a circular conduit has the least surrounding surface than any other shape of conduit. As a result, all else being equal, a circular duct offers less resistance than an oval or elliptical duct. Similarly, a square duct will resist flow less than a rectangular duct. Since resistance must be overcome by the fan motor, when energy consumption is a consideration, the round or square duct is preferred to those ducts with elongated cross-sections. However, this must be balanced with the amount of space and vertical dimension of the duct. Generally, to minimize the vertical space that is dedicated to air distribution, the architect would like the cross-section of the ducts to be as low as possible. This is particularly important for high-rise buildings, where reducing the height of the ducts by just six inches can result in significant reductions in the overall height of the building or offer the opportunity to add more occupied (rentable) floors. However, as the aspect ratio – i.e., the ratio of width to height of the ducts – grows larger, the surface-to-volume ratio increases resulting in greater friction losses throughout the duct runs, requiring a larger fan motor to push the air through the system. The same may be true when integrating the air distribution into walls. Because walls tend to be thin, high aspect ratios can allow ducts to be placed in the wall but with the same consequence for friction losses as previously noted. Using ducts with high aspect ratios can significantly increase the energy use throughout the life of the building.

The inlet conditions and entry type also affect the resistance to airflow. An open-ended (plain) or flanged entry tends to create turbulence downstream of the entry. This disturbs the stream, resulting in loss of energy. As previously mentioned, a bell mouth entry, on the other hand, creates a smooth airflow pattern with less turbulence (Figure 7.27). Once again, if energy consumption is a consideration, then the entry condition should be considered.

For sharp-cornered entry configurations (see Figure 7.7), the loss coefficient k needed in Equation (7.5) is equal to 0.5. If the entry mouth is well rounded, the coefficient is only 0.03. On the other hand, for exit configurations (see Figure 7.2) – i.e., pipes or ducts discharging flow in a wide-open space – the coefficient is equal to 1. This means that all the kinetic energy of the flow is lost. In fact, regardless of the shape of the exit, it is not possible to recover any of the kinetic energy.

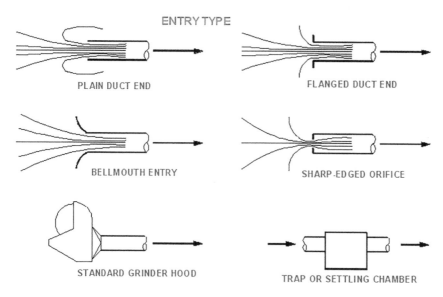

Figure 7.27 Types of duct entry conditions, source: author's figure

In addition to the resistance to airflow from the shape of the duct, three other air distribution issues should be considered. First is the length of the duct. The longer the duct, the greater the resistance. In fact, friction losses are directly proportional to the length of the duct, as indicated in Equation (7.5), and therefore, the architect should work with the mechanical engineer to keep the duct length to a minimum. This is a key space-planning consideration when locating mechanical rooms.

Secondly, turns and bends in the duct create friction and resistance to airflow. As previously introduced, any fluid in motion like air or water has momentum and does not "like" to make sharp turns. Abrupt turns in a duct create turbulence, which adds to the losses, reduces the stream energy, and creates noise. Losses along turns can be expressed in Equation (7.5), and minor loss coefficients are shown in Figure 7.28. Turns in ducts should be as gradual as possible rather than abrupt. For this, turning vanes are often inserted into the duct to smoothly turn the air, thus reducing turbulence and noise (Figure 7.29). It is important that the turning vanes be aligned to direct the flow appropriately.

Thirdly, changes in shape along the duct should be avoided for the same reasons as noted earlier. Changing the duct shape disturbs the airflow stream, which reduces the effectiveness of air distribution. The loss coefficients of sudden expansions or contractions of a duct are listed in Figure 7.30. If changes in duct geometry are required, then a smooth transition should be used between the two shapes. It should be noted that diameter changes in the ductwork are common when the volume of air changes due to branching or discharge of some of the air.

Another configuration of significance in HVAC ducting systems is bifurcations – i.e., fittings that direct that flow into two or more branches. Ideally, these bifurcations should not involve any turning of the flow, but this is not easy, and most configurations have the form of a tee. The sharp turn into the branch of the tee induces large losses, and the corresponding coefficient is 1 (see Figure 7.31). The coefficient for the line branch is just

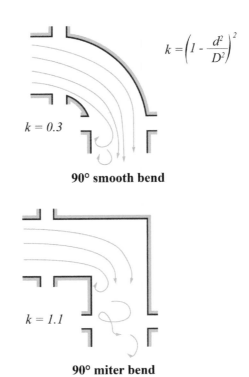

$$k = \left(1 - \frac{d^2}{D^2}\right)^2$$

$k = 0.3$

90° smooth bend

$k = 1.1$

90° miter bend

Figure 7.28 Minor loss coefficients in bends, source: author's figure

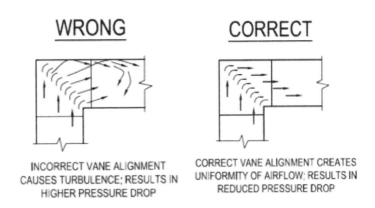

WRONG

CORRECT

INCORRECT VANE ALIGNMENT
CAUSES TURBULENCE; RESULTS IN
HIGHER PRESSURE DROP

CORRECT VANE ALIGNMENT CREATES
UNIFORMITY OF AIRFLOW; RESULTS IN
REDUCED PRESSURE DROP

Figure 7.29 Typical duct turning vanes, source: author's figure

0.2. Of greater significance is the fact that the flow into the branch is severely reduced. The distribution in the two lines also depends on the load that the flow encounters further downstream. The flow avoids the line that leads to many obstructions, or more tee bifurcations. In an HVAC system, the rooms or spaces served by the branch line do not receive adequate conditioned air.

Once the air is distributed through the duct system to the occupied space, the distribution of the air in the room should be considered. Generally, there are four types of room air distribution, *(a) mixed, (b) short-circuit, (c) displacement, and (d) piston flow.* The

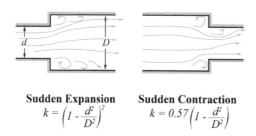

Sudden Expansion
$$k = \left(1 - \frac{d^2}{D^2}\right)^2$$

Sudden Contraction
$$k = 0.57\left(1 - \frac{d^2}{D^2}\right)$$

Figure 7.30 Minor losses in expansions and contractions, source: author's figure

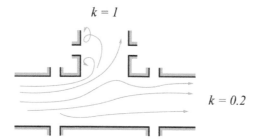

$k = 1$

$k = 0.2$

Figure 7.31 Minor losses along bifurcation, source: author's figure

air distribution to nonresidential spaces is an important consideration as fresh ventilation air is required in the occupied zone, and it must be distributed as evenly as possible. Generally, the most common room air distribution approach is mixed ventilation. For this case, the ventilation air is typically supplied to the room through diffusers located in the ceiling and returned through the ceiling. As can be seen in Figure 7.32 top, the ventilation air mixes with the air already in the room. The consequence is that the air that is breathed is a mixture of fresh ventilation air and potentially stale room air. When indoor air quality is a primary concern, the mixed ventilation approach is not ideal. For the short-circuit distribution, as described earlier, the situation is even worse, as the fresh ventilation air is short-circuited to the return inlet, and does not reach the breathing zone. This is the least effective ventilation situation, Figure 7.14. Short-circuiting can occur when the velocity of the air through the supply air diffuser is low and lacks the necessary momentum to distribute to the breathing zone. This can be the case when VAV systems are operating at low flow. Diffusers specifically designed for VAV systems are available to avoid this problem.

Due to the increased concerns for indoor air quality over the past few decades, two alternative ventilation distribution patterns were introduced: displacement ventilation and piston flow ventilation. Displacement ventilation typically introduces air to the room through diffusers located in or near the floor rather than the ceiling (mixed) and returns the air through or near the ceiling. The piston flow is a variation of this flow pattern, but for the piston flow, the air is distributed more uniformly through multiple inlets in the floor.

There are at least three potential benefits to the displacement or piston flow systems when compared to the mixed ventilation. First, the ventilation effectiveness is very high since the fresh ventilation air is being supplied directly to the occupied breathing zone.

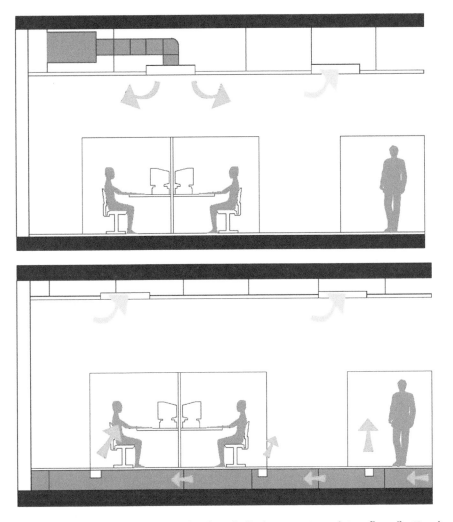

Figure 7.32 Mixed or short-circuit (top) and displacement or piston flow (bottom), source: author's figure

In addition, unlike the mixed strategy, for the displacement and piston flow, the supply air pushes the stale room air upward toward the return, and, consequently, the air being breathed is fresher.

Second, since the displacement and piston flow patterns are vertical from the floor to the ceiling, heat from lights, people, and equipment in the room is more effectively removed. Because this heat is removed and not mixed with the room air, the cooling loads in the space are less than for mixed flow, which can result in lower energy consumption when compared to mixed ventilation. An important design consideration for the displacement or piston flow systems is the need for underfloor distribution. This may mean more space in the section for the floor and greater floor-to-floor height.

Thirdly, for the displacement and more so for the piston flow, the air is distributed to the room through larger openings than the diffusers for mixed flow. Because the inlet area is greater, the flow velocity is lower (given equal volume), and the noise from the air distribution is less. This can be important when low noise criteria ratings are desired for rooms such as classrooms and performance spaces.

As with wind or thermal buoyancy–driven airflow, mechanically driven airflow can have several impacts on the building design that should be considered early in the design.

References

[1] ASHRAE (2010) *ANSI/ASHRAE Standard 62.1–2010, "Ventilation for Acceptable Indoor Air Quality."* Atlanta GA, The American Society of Heating, Refrigerating and Air-Conditioning Engineers, Inc., p. 5.
[2] ASHRAE (2012) *Fundamentals – HVAC Systems and Equipment*. Atlanta GA, The American Society of Heating, Refrigerating and Air-conditioning Engineers, Inc., pp. 20.4, 20.6.
[3] NEBB (2022) *Building Performance*. Available from: www.nebb.org/assets/1/7/NEBB_Talk_I_Basics.pdf [Accessed December 2019].
[4] White, F.M. (2021) *Fluid Mechanics*, 9th edition. New York, NY, McGraw Hill.

Chapter 8

Measurement Instrumentation

In architectural practice, as well as in design research, there are many opportunities when the measurement of wind effects on buildings may be informative. This might be to understand and quantify pressure distributions on walls or roofs or resistance of the structure to forces from high wind conditions such as hurricanes and to map out internal air streams that affect natural ventilation. These measurements may be on or around as-built structures exposed to ambient wind conditions or with scale models in wind tunnels. Model studies in wind tunnels are fairly common and have been used by such notable architectural engineering firms as BuroHappold.

The behavior of the flow over or through structures can be documented experimentally in full-scale or in scaled-down models. Flow properties can be measured or visualized with instrumentation. There is a large number of instruments available to measure physical properties of flow. Instruments are available to measure flow velocity, wall pressure, flow rate, fluid temperature, and other flow properties. Methods are also available to capture the global features of the flow. These methods and instruments can be used in the laboratory for flows over models or in full scale for flows over structures.

Instrumentation for the measurement of flow velocity varies in complexity and accuracy, as well as in the ease of use. Their price also varies from a few thousand dollars to many hundreds of thousands of dollars. Most instruments can make point measurements – i.e., they can measure flow properties at a point in the flow. To provide a broader view of the flow, these instruments must be traversed from point to point, a process that is often automated. But instrumentation is available to capture a domain of the flow with one stroke. Most useful and easy to use are flow visualization methods, like smoke visualization in a wind tunnel or dye in a water tunnel or a laminar flow table (Figure 8.22). Techniques have

DOI: 10.4324/9781003167761-8

been developed to actually measure velocity along a plane, most prominent among these methods is Particle Image Velocimetry.

Architects and construction engineers are most interested in measuring pressure over a solid surface, like a building wall, a roof, or an internal wall of a duct. To measure the pressure at a certain point on a wall, a small hole is drilled at that point, a solid pipette is press-fit on the other side of the wall, a flexible tube is attached to the pipette, and then connected to a pressure gauge as shown in Figure 8.1.

The hole, often referred to as a pressure tap, should be drilled close to perpendicular to the wall, but a small inclination will not affect the reading. Moreover, the pressure reading will not be affected by a fully developed boundary layer, turbulence, or even small eddies at the opening of the pressure tap (Figure 8.2).

To measure the pressure distribution over a surface, many pressure taps are drilled and the tubes are collected and attached to pressure manometers in the back, as shown in Figure 8.3.

The easiest and least expensive method to measure pressure is the tube manometer. This is just a U-tube filled with a liquid of known specific weight. It can be constructed in a few minutes with flexible tubes and can be filled with colored water. The manometer

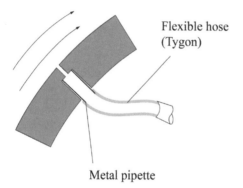

Flexible hose
(Tygon)

Metal pipette

Figure 8.1 Pressure tap and flexible tubing connection, source: author's figure

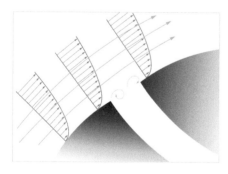

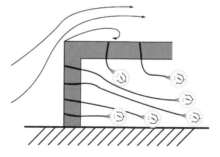

Figure 8.2 The opening of a pressure tap, source: author's figure

Figure 8.3 Pressure measurements over a surface of a structure, source: author's figure

shown in Figure 8.4a measures the pressure difference between the pressure p_1 in the tank on the left and the pressure p_2 in the tank on the right. The pressure difference then is given by

$$p_1 - p_2 = \gamma h,$$ (8.1)

where γ is the specific gravity of the liquid in the manometer, and h is the difference in the elevation of the liquid in the U-tube. The manometer in Figure 8.4b measures the pressure difference between the pressure in the tank p_1 and the atmospheric pressure, p_a. The measured quantity h is related directly to the gauge pressure in the tank. Figure 8.4c shows a commercially available U-tube manometer.

This simple instrument can be used in practice to provide useful information. For example, it can monitor the pressure in an HVAC system (Figure 8.5) by recording the pressure p_1 at the discharge of the blower; the pressure after a tee branch, p_2; or the pressure after a number of bends, p_3.

More sophisticated instruments to measure pressure are available. Such instruments provide an electronic reading, which can be interfaced with a computer to record pressures over a certain time period. Most common among these instruments are pressure scanners, which consist of many piezoelectric pressure gauges packed together in a box (Figure 8.6). The pressure tubes from the pressure taps are then connected directly to the pressure scanner.

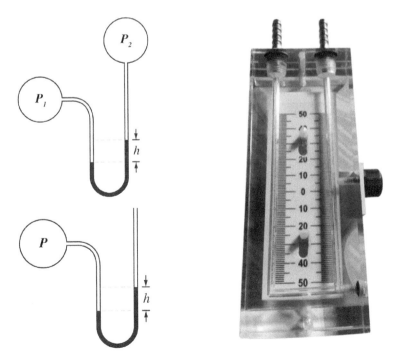

Figure 8.4 The U-tube manometer, source: author's figure

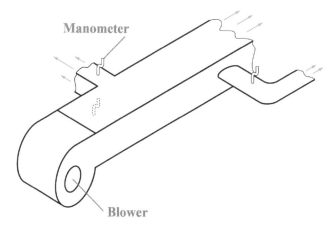

Figure 8.5 U-tube manometers can measure pressure along a duct system, source: author's figure

Figure 8.6 Pressure scanner (courtesy of Scanivalve)

The least expensive velocity measuring tool and the easiest to use is the Pitot tube. This is just a small metal tube with one end open and the other connected to a manometer. The Pitot tube must be aligned with the flow, as shown in Figure 8.7, but a misalignment of as much as 15° has very little effect on the measurement.

The pressure measured by a Pitot tube records the dynamic pressure p_o introduced in Chapter 4:

$$p_o = p_\infty + \rho V_\infty^2 / 2 .\tag{8.2}$$

The quantities with the subscript ∞ should be pressure and velocity far upstream. But in practice, these quantities could be a few tube diameters upstream of the tip of the instrument. Thus, they could be the local pressure, p, and velocity V of the flow:

$$p_o = p + \rho V^2 / 2 .\tag{8.3}$$

The velocity at the point of interest then is given by

$$V = [2(p_o - p)/\rho]^{1/2} .\tag{8.4}$$

Simple Pitot
tube

Static
source

Pitot - static
tube

Figure 8.7 Pitot tubes, source: Wikimedia

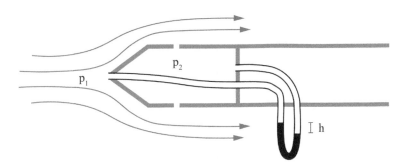

p_2

p_1

h

Figure 8.8 A Pitot-Static tube, source: author's figure

To complete the measurement, one needs the local static pressure, p. In case the measurement is made inside a duct, then the pressure measured on the wall of the duct is a good approximation of the local static pressure. But a simple modification of the Pitot tube is the Pitot-Static, which can actually measure the local static pressure as well. To this end, a separated compartment is provided, and small holes are opened along a circle downstream of the tip, called the static ring. (Figure 8.8). The average pressure in this compartment is a very good approximation of the static pressure in the immediate neighborhood of the tip.

The tip of a Pitot-Static could be conical or hemispherical. A Pitot-Static with a hemispherical tip is shown in Figure 8.9. The static ring on this probe has eight static taps. Pitot-Static, often referred as just Pitot probes are commercially available at very reasonable prices.

A Pitot-Static tube is easy to use, but its accuracy decreases with the incident angle. The user should know the direction of the oncoming flow in the vicinity of the point of interest. A different design of a tube probe that eliminates the need to align the probe with the flow and can actually provide measurements of both the magnitude and the direction of the flow at the point of measurement is the multi-hole probe.

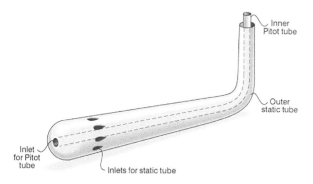

Figure 8.9 A Pitot-Static with an eight-hole static ring, source: ScienceDirect

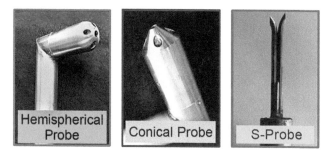

Figure 8.10 Hemispherical, conical, and hemispherical five-hole probe tips, source: NIST

Multi-hole probes look very similar to a Pitot-Static probe, but along their tips, they have more than one hole. Most common among them are the five-hole or the seven-hole probe. The tip of these probes can be conical, hemispherical, or S-probe, as shown in Figure 8.10.

The holes are internally connected to pressure gauges, and the pressures along the tip pressure taps are recorded. A multi-hole probe must be calibrated in a small calibration wind tunnel, and with proper instrumentation, it can read the magnitude and the direction of the velocity, as well as the static and dynamic pressure at the point of measurement. Multi-hole probes are commercially available.

More sophisticated and accurate instruments are available, but they are more expensive and much harder to use. Most of them are also limited in their physical domain of use and therefore are not appropriate for full-scale measurements. The most common such instruments are the hot-wire anemometer, the laser-Doppler velocimeter (LDV), and the ultrasonic anemometer. These three instruments measure again the velocity at one point in space.

The probe of a hot-wire anemometer consists of two fine needles the size of sewing needles with a very thin wire connecting their tips. An electronic system passes an electrical current through the wire which heats it. When the probe is inserted in the flow, the sensing wire is cooled. The electronic components of the instrument generate the proper current level to keep the temperature of the wire constant and measure the voltage necessary to sustain the temperature at the chosen temperature level. The instrument is calibrated to convert the voltage to flow velocity. This is the most common hot-wire anemometer, and is known as the constant-temperature anemometer (CTA). This instrument measures the velocity normal to the wire. More complex tips with multiple stretched wires are available to measure all three velocity components (Figure 8.11).

The LDV is based on an entirely different principle. Two laser beams are directed to cross each other at the point of measurement in the flow. In the space of the beam intersection, optical fringes are formed, which can be described in a simple way as planes of darkness and light. These planes are formed within an oblong special domain, which is known as the measurement volume of the LDV. This volume is very small, usually the order of a fraction of a millimeter. A particle that passes across these planes therefore can be seen lighted when between two planes of darkness, and as it moves fast across the light planes, it emits an on-and-off light signal. The faster the speed of the particle, the higher the frequency of the fluctuation of the light signal. The processing of the signal requires the measurement of the frequency of the emitted signal, and with the spacing of the fringes known, this can be converted to the velocity of the particle.

LDV measures the velocity of particles that drift with the flow, not the velocity of the fluid. This requires the seeding of the flow with very small particles that follow the flow. These particles must be small so that they move together with their neighbor fluid elements, and that they do not interfere and thus alter the flow properties. A large number of particles is required to ensure that a few particles are always within the measuring volume.

Figure 8.11 Hot-wire anemometer probe (courtesy of Dantec Dynamics)

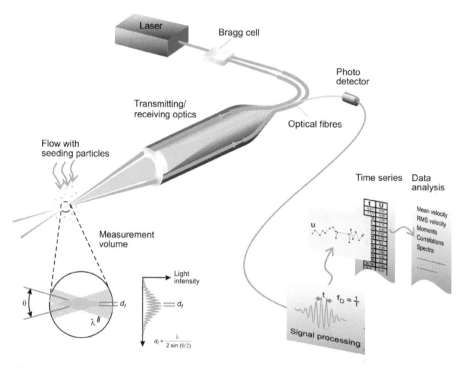

Figure 8.12 A compact LDV probe (courtesy of Dantec Dynamics)

An LDV system requires the sending optics, which consists of laser beams and lenses to form beam intersections, and receiving optics, which consists of lenses and light sensing equipment. The signal emitted by the particles is extremely weak, and its sensing requires special instruments known as photomultipliers. The optical signal is then converted to flow velocity at the point measurement. Modern LDV equipment is available today that combines the sending and receiving components in one stand-alone probe, as shown in Figure 8.12.

LDV systems measure the component of the velocity normal to the axis of the probe and within the plane of the two laser beams. LDV probes are available that send four laser beams along two planes perpendicular to each other, and thus they can measure two components of the velocity.

LDV probes are somewhat constrained in the distance they can reach in the flow. Their use is therefore limited to flows over small, scaled-down models. Moreover, other serious limitations of LDV are the need to seed the flow with small particles and the need to have optical access to the domain of interest. The latter limitation dictates that wind tunnel walls are transparent to allow laser beams to go through. In Figure 8.13, we show an LDV probe direct team laser beams through a test section transparent wall, to measure flow velocity over a model car placed in the test section of a wind tunnel.

LDV can resolve high-frequency fluctuations of the velocity. They can therefore record turbulent fluctuations. Multi-hole probes are limited to frequencies of about 100 Hz, while CTA and LDV can measure velocity fluctuations as high as a few kHz. CTA and LDV are very well suited for resolving small-scale turbulence and are therefore very useful in

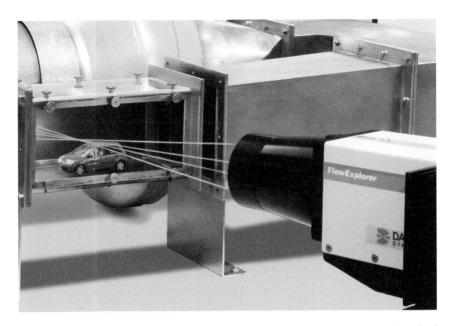

Figure 8.13 LDV probe sending laser beams in a test section to measure the velocity over a model (courtesy of Dantec Dynamics)

basic fluid mechanics research. But this type of temporal resolution is not necessary for the study of flows over full-scale structures, or even flows inside enclosures, like the interior space of buildings.

Wind Sensors

Before the design of a structure begins, information should be collected on the weather conditions in the area, especially on the site where the structure will be erected. Information on temperature variations, precipitation patterns, and, of course, wind directions and magnitude can be provided by public or private weather monitor stations. But it is a good practice for the designer to obtain some information on the spot, using special instrumentation.

The instruments described in the previous section can be used to measure wind velocity, but they are sensitive to humidity and temperature variations. Instruments that are immune to negative weather conditions are the cup anemometer and the ultrasonic anemometer. These instruments can be mounted on posts or towers, and they can record conditions over long periods of time.

The most common instrument to measure the wind velocity is the cup anemometer shown in Figure 8.14. This instrument consists of a rotor – i.e., a rotating set of spokes with hemispherical or conical cups attached to the spoke ends. The principle of its operation is the fact that the side of a cup that exposes a hollow surface to the wind generates larger forces than the side that presents to the wind the smooth curved hemispherical wall. As a result, the system tends to turn so that the hollow-cup sides retreat, and allow the other

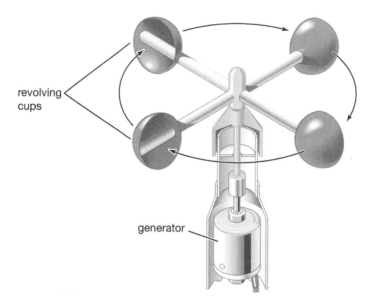

revolving
cups

generator

Figure 8.14 A cup anemometer, source: Wind and Weather Tools

side to advance against the wind. The frequency of rotation can be calibrated to return the wind velocity. This can be done with a generator.

The cup anemometer is insensitive to the direction of the wind. It just measures its magnitude. In fact, if there is a component of the wind in the vertical direction, then this instrument returns only the horizontal component of the wind. The cup anemometer has a very poor frequency response. It cannot measure fluctuations of the wind, which are quite important if atmospheric turbulence is expected to be recorded. This is because the rotor attains rotational momentum, and it cannot slow down or accelerate fast enough to record lower or higher speeds.

The direction of the wind is very important for the design of structures, perhaps more important than the magnitude of the wind. To record the direction of the wind, a vane probe can be added, as shown in Figure 8.15. With a probe like that, the history of the wind magnitude and direction can be recorded for long periods of time.

A more accurate probe and one that can respond well to atmospheric turbulence is the ultrasonic anemometer. This instrument consists of pairs of transducers, which can emit and receive ultrasonic pulses. The transducers are mounted at the tips of brackets, as shown in Figure 8.16. Let us assume now the line along two opposing transducers is along the north-south direction, and the other two are along the west-east direction. The principle of operation is based on the fact that sound travels with the speed of sound through a medium at rest. But if wind is blowing, sound travels faster in the direction of the wind and slower in the opposite direction. Sound is literally carried by the moving medium. An ultrasonic anemometer measures the time taken for an ultrasonic pulse of sound to travel from one transducer, say the north transducer to the transducer on the opposite side, the south transducer. The signal processing instrumentation compares the time for a pulse to travel from north to south to the time for the pulse to travel in the opposite direction. For a wind blowing from the north, the faster the wind, the shorter the time required for the

Figure 8.15 A cup anemometer and a vane probe, source: Learning Weather

Figure 8.16 A 3D Windmaster ultrasonic anemometer, source: Wikimedia

signal to travel from north to south and the slower to travel from south to north. This information is converted to the velocity component along the north-south direction. Two more transducers along a line perpendicular to the line north-south provide the other component of the velocity – i.e., the east-west direction.

Geometric and Dynamic Similarity

Scaled-down structure models can be tested in facilities like wind tunnels or water tunnels. Geometric scaling is a common and well-understood practice of architects and engineers. But scaling down a structure does not guarantee that the flow over it will be geometrically similar to the flow over the prototype. A typical example that indicates the wide range of flow patterns over the same basic object is the flow over the simplest geometric configuration – i.e., a circular cylinder. The flow patterns over a circular cylinder vary wildly, as shown in Figure 8.17. These patterns depend on the Reynolds number, Re, defined in Chapter 3, and further discussed in this section.

Re,<100; Frame 3, 100<Re< 1000; Frame 4, 1000<Re<50,000; Frame 5, Re.600,000

It is important to emphasize here that these are patterns of flow over circular cylinders at different speeds or cylinder dimensions in uniform flow, and yet the flow patterns that develop are drastically different from each other. It is even more interesting that with a steady free stream, and a fixed model, some unsteadiness in the flow may develop as described earlier in our discussion of vortex shedding. The patterns that develop over a circular cylinder and in fact over different bodies depend on the free-stream velocity V, a typical length of the body, and the kinematic viscosity of the fluid, ν. In fact, the flow patterns depend on the combination of these parameters in the form of the Reynolds number, which is defined in the next paragraph. For example, the pattern shown in Frame 2 of Figure 8.17 could be

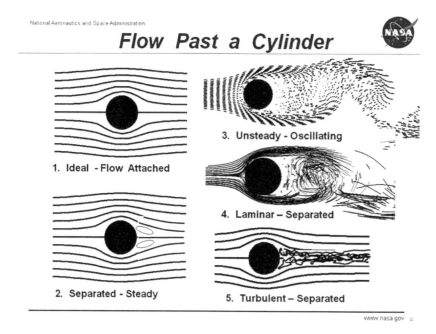

Figure 8.17 Flow over a circular cylinder for different Reynolds numbers, source: Wikimedia

the flow of honey at a speed of 0.01 ft/s over a circular rod with a diameter of 2 inches, or the flow of air at a speed of 1 ft/s over a sewing needle of 1/32 inches in diameter. Both these flow situations have the same Reynolds number – namely, a Reynolds number of 10. Frame 1 of this figure shows the pattern calculated for ideal flow – i.e., flow of an inviscid fluid – but it is very close to patterns of flow at Reynolds numbers less than 1. The patterns of Frames 4 and 5 we discussed earlier in Chapter 3, see also Figure 3.28.

When tests are conducted with scaled-down models of buildings, it is important to arrange for the Reynolds number of the model flow to match the Reynolds number of the prototype flow. Since these tests are most of the time conducted in air, it is very difficult to match the Reynolds numbers. In simple terms, one cannot run wind tunnel tests with very small building models. The reasoning for this fact will be discussed in the section.

In aerodynamic testing, there are special scaling laws that ensure that flow patterns over a model, like streamlines, forces, and pressures will be similar to those over the prototype. The similarity of flow patterns and force distributions and forces is called dynamic similarity. For flows with wind speeds less than about 200 mph, dynamic similarity requires that the Reynolds number of the model flow is equal to the Reynolds number of the prototype. The Reynolds number is defined as follows:

$$Re = \rho VL/\mu . \qquad (8.5)$$

Here ρ and μ are the density and viscosity of the fluid, V is the velocity of the fluid flowing around the model, and L is just a typical dimension of the object. For example, as discussed earlier, it could be the diameter of a cylinder. In experiments with model buildings, L could be the height of a building. The ratio of viscosity, μ over density ρ is the kinematic viscosity, ν, and the Reynolds number then becomes

$$Re = VL/\nu . \qquad (8.6)$$

The laws of dynamic similarity are discussed in all textbooks of fluid mechanics, as, for example, in White [1]. In the case of airflow over structures, the law of dynamic similarity dictates that the Reynolds number of the model flow be equal to the Reynolds number of the prototype.

$$V_m L_m / \nu_m = V_p L_p / \nu_p \qquad (8.7)$$

Here the subscript m denotes the model, and the subscript p denotes the prototype. It should be noted that each of these quantities can be expressed in any unit, provided that the same unit is used for each quantity on the two sides of the equation. For example, the velocities could be expressed in miles per hour and the lengths in centimeters.

If testing is being done with the medium of air in both the model and the prototype, in other words, if we model the flow of air over a prototype, and we run our experiments in a wind tunnel, then $\nu_p = \nu_m$. The kinematic viscosity drops out of the law of dynamic similarity.

$$V_m L_m = V_p L_p , \qquad (8.8)$$

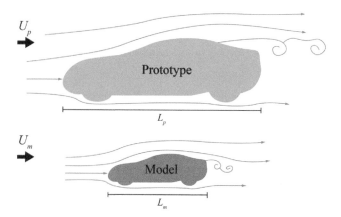

Figure 8.18 A prototype and a scaled-down model, source: author's figure

which can be rewritten as

$$V_p / V_m = L_m / L_p = \lambda . \tag{8.9}$$

Here, λ is the geometrical scale. This law presents perhaps an unexpected requirement for model testing. If a model is scaled down for a test, then the speeds of the test should be scaled not also down, but up.

A numerical example will make this rule of dynamic similarity easier to comprehend (see Figure 8.18). Let us assume we need to design a wind tunnel test of an automobile of length L_p = 20 ft at a speed of V_p = 70 mph. For a model scaled down by a scale of λ = 1/2 – i.e., a model of L_m = 10 ft – the velocity ratio should be $V_p / V_m = \lambda$ = 1/2, and thus, $V_m = 2V_p$ = 140 mph. Simply stated, if the model is scaled down by a factor of ½, then the test speed should be twice as large as the prototype.

There is a serious limitation for wind tunnel testing based on the dynamic similarity rule of $Re_m = Re_p$. The basic assumption for this rule to guarantee similarity is that both model and prototype flows do not exceed a speed of about 200 mph. For higher speeds, a new set of flow phenomena dominate the flow, and dynamic similarity does not hold. This means that there is a limitation to scaling down models. In the numerical example described in the previous paragraph, we cannot use a model of 5 ft in length because our dynamic similarity will indicate a model test speed of 280 mph, which exceeds the limit of 200 mph.

This limitation indicates that it is not possible to test in wind tunnel models of automobiles or airplanes. For this reason, all the major automobile manufacturers conduct tests in very large wind tunnels with full-scale models. And since airplanes are much larger than the test section of any wind tunnel, testing in full-scale can be made only of components of an airplane like a section of the wing. Testing airplanes in wind tunnels can actually be handled only if scaling rules can be adjusted and corrections made.

Another unexpected result of the dynamic similarity of Equation (8.7) indicated by engineering analysis is that forces exerted on the prototype are equal to forces exerted on the model, $F_p = F_m$. For example, scaling down a wing to half the original size, one must run the model test at twice the prototype speed, but then the forces measured on the model, like the lift the wing produces are equal to the forces on the prototype. We can now estimate easily how pressures scale on the prototype and the model. Since pressure is a force over an area, typical pressures could be the ratio of a force over a typical area so that pressure measured on the model can be expressed as $p_m = F_m/L_m^2$, and thus with $L_m/L_p = \lambda$, the scale ratio, we have

$$p_m = F_m/L_m^2 = F_p/\lambda^2 L_p^2 = p_p/\lambda^2 . \qquad (8.10)$$

This implies that pressures measured on a model half the size of the prototype are four times larger than the pressures on the prototype.

Aerodynamic Testing of Structures

Aerodynamic testing can provide information that can be useful in the design of a structure, or for buildings that are already constructed. Testing can provide technical information on developing pressure distributions, on dynamic flow fluctuations, or flow patterns that interfere with the comfort of the occupants.

Wind tunnels are facilities that create flow conditions in a test section similar to the prototype situation. Testing conditions for vehicles are different than the conditions required for buildings. Testing for vehicles must be carried out in a free stream. A proper free stream should consist of a uniform velocity across the entire area of the test section. Uniformity implies that the velocity should be the same in magnitude as well as in direction across the entire cross-section at the entrance of the test section. Moreover, the free stream should be free of turbulence. These are the conditions experienced by a vehicle in motion, a car, or an airplane. This is because the aerodynamics of an object moving in still air is the same as the aerodynamics of an object being fixed on the ground with a stream of air flowing over it.

The test section conditions for the testing of fixed structures are different. To simulate the flow over a structure in a wind tunnel, the velocity should have the profile of the atmospheric boundary layer, and it should also include the turbulence of the lower part of the boundary layer.

A typical wind tunnel is shown in Figure 8.19. The air in the wind tunnel is set in motion by a fan. The flow is guided through a closed circuit. Turning vanes are installed along the 90° turns to prevent the flow from separating and creating large recirculating regions. Turbulence is generated along the passages through turning vanes or across the blades of the driving fan. All ducts are effectively diffusers – i.e., they are designed with mild area expansions – which induces a lowering of the velocity all along the closed circuit, without introducing any turbulence. This allows some of the turbulence to be reduced and leads to the largest section right upstream of the test section, the settling chamber. To reduce any irregularities in the uniformity of the flow, a honeycomb is included upstream of the test section. Fine screens are installed that further reduce the free-stream turbulence. A control room is provided around the test section so that the operators can insert models and instrumentation.

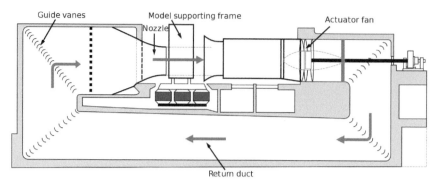

Guide vanes   Model supporting frame   Actuator fan
Nozzle

Return duct

Figure 8.19 A closed circuit wind tunnel, source: Wikimedia

Within the limitation discussed earlier, a wind tunnel for studying the effects of winds on buildings had been in operation at the College of Architecture and Urban Studies at Virginia Tech (recently destroyed in a fire). This wind tunnel was used in many studies such as that by Arron West to study the effects of wind direction on natural ventilation flow for the addition to the Amsterdam World Trade Center by Kohn, Pedersen, and Fox, as shown in Figures 8.20 and 8.21.

While a wind tunnel is relatively expensive to construct and instrument, and occupies a significant amount of room, a viable alternative for teaching principles of fluid flow and for studying relatively simple cases of fluid motion is a laminar flow table. Shown in Figure 8.22, a laminar flow table is essentially a slightly tilted water table where water flows downhill in laminar motion from an inlet on one side to an outlet on the other side. The flow velocity can be controlled, within limits, through a valve that varies the water flow rate to the inlets. To visualize the 2D flow pattern a dye is injected at the inlets to create flow streams. The laminar flow table can be used to visualize flow patterns through and over 2D objects, to see changes in the flow and to identify locations with eddies and turbulent flow. Within the limitations of not matching Reynolds numbers with the prototype and the constraints of two-dimensional flow, the laminar flow table can be an effective instrument for improving one's understanding of fluid flow.

Wind tunnel test sections are available in a great range of sizes, from sections as small as a foot by a foot to 20' or 30' x 50'. The price of these tunnels varies according to the size. Moreover, the energy required to drive the fan of a very large wind tunnel is considerable. The energy needed to operate a few of the largest wind tunnels at NASA laboratories is close to the energy required to support a small village.

Testing the aerodynamics of architectural structures in wind tunnels poses similar limitations to testing vehicles like automobiles or airplanes. The most significant limitation is the need to match the Reynolds number of the prototype design and the model. In a nutshell, this means that vehicles or buildings must be tested practically in full scale. Historically, aerospace engineers limited their experiments to components of their prototype design, like a section of a wing, or the nose of an aircraft. This option is of course available

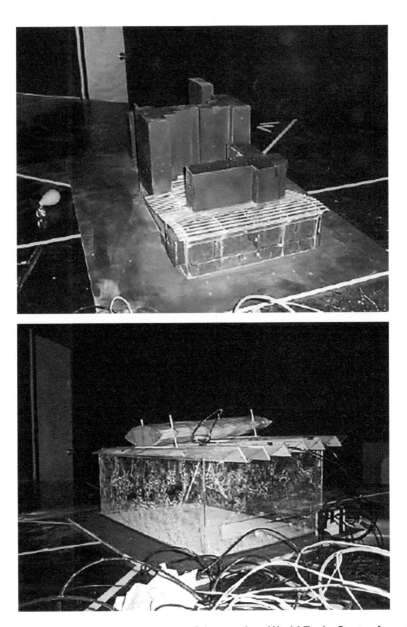

Figures 8.20 and 8.21 Scale models of Amsterdam World Trade Center for wind tunnel testing, source: author's figure

to architects who could test only a part of the designed structure, like the design of a roof overhang. Testing of prototypes in full scale requires the use of very large tunnels, which is very expensive, and even this option is not possible for large prototypes like a commercial airline vessel, or a large multi-story building. But an alternative testing solution is available for architects. They can conduct testing in full scale in open, outdoor terrain and in the

Figure 8.22 Laminar flow table images, source: Armfield

presence of strong atmospheric winds. This advantage may be somewhat mute because it implies that the prototype is already constructed.

Modeling the aerodynamics of buildings imposes difficulties not encountered by aerospace engineers. Modeling and testing the aerodynamic response of buildings requires modeling of the atmospheric boundary layer. In aerodynamic testing of vehicles, the oncoming stream should be uniform. But to model the aerodynamics of buildings and other fixed structures, the atmospheric boundary layer must be created in the test section of the wind tunnel. In other words, the wind tunnel test section stream should include a boundary layer – i.e., velocity profiles that decrease as the ground is approached. Moreover, special care is required to generate models of atmospheric turbulence. This is a strange requirement because all wind tunnels are designed with the best available techniques to reduce turbulence as much as possible, and here architectural testing requires the introduction of turbulence. But the important fact here is the need to introduce the specific character of atmospheric turbulence. This requires careful measurements of atmospheric conditions in different terrains and then preparing the necessary inserts to create these conditions in the test section of a wind tunnel.

Marshal [2] and Tieleman et al. [3] conducted tests in the Virginia Tech wind tunnel, a facility with a test section cross-section of 183 m x 1.83 m and a length of 7.5 m. They introduced vertical spires at the entrance to the test section with widths decreasing with height, as shown in Figure 8.23. With blockage of the stream increasing as the ground is approached, the velocity also decreases to simulate an atmospheric boundary layer. The blockage of the spires also introduces some turbulence. To simulate the proper characteristics of atmospheric turbulence, it was also necessary to place on

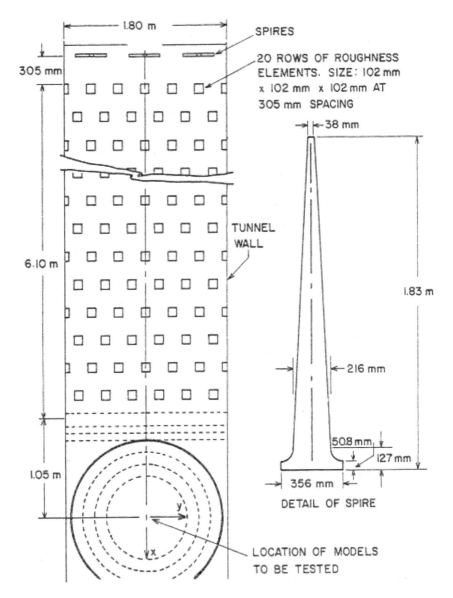

Figure 8.23 Inserts in a wind tunnel test section that simulates atmospheric turbulence, source: Marshal [2] and Tieleman et al. [3]

the floor of the test section rows of small cubic elements (10.2 cm in height) as shown in Figure 8.23.

These inserts generated a velocity profile at about a distance of 6 m from the entrance to the test section, which simulates the lower part of an atmospheric boundary layer (Figure 8.24) In this figure, velocities and lengths are ratios to reference values indicated in the caption.

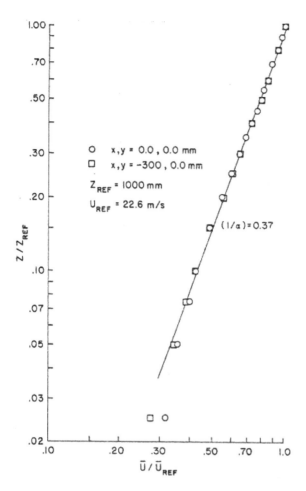

Figure 8.24 Velocity profile in the Virginia Tech wind tunnel test section, source: Marshal [2] and Tieleman et al. [3]

Tieleman et al. [3] and later Reinhold et al. [4] placed building models on a turntable positioned about 7 m from the entrance to the test section. The turntable is shown in Figure 8.20 as a circle. It is actually a circular plate that can be rotated to change the angle with which the stream is approaching the model.

Full-scale testing can be done in open terrain. But this requires a space not obstructed by other buildings or by trees and other vegetation. Moreover, it is necessary to document with great care the features of the oncoming wind stream.

A field laboratory has been constructed in Lubbock, Texas, by researchers at Texas Tech University [5, 6]. This facility consists of a meteorological tower and a test structure shown in Figure 8.25. The aim is to instrument a full-scale structure named the

Figure 8.25 The Texas Tech University field facility for full-scale testing, source: TTU

Wind Engineering Research Field Laboratory test structure in order to measure pressure distributions on all sides when the structure is exposed to the wind. It is necessary to have detailed information on the direction, the magnitude, and the turbulent character of the wind that impinges on the test structure. This was accomplished by an instrumented tower with sensors at elevations of 8 ft., 13 ft., 33 ft., 70 ft., and 160 ft. The test structure is a rectangular, flat-roofed building instrumented with 240 pressure taps on most of its surfaces.

References

[1] White, F.M. (2021) *Fluid Mechanics*, 9th edition. New York, NY, McGraw Hill.

[2] Marshall, R.D. (1978) "A Study of Wind Pressures on a Single-Family Dwelling in Model and Full-Screen," *Journal of Industrial Aerodynamics* 1: 177–199.

[3] Tieleman, H.W., T.A. Reinhold and R.D. Marshall (1977) "On the Wind-Tunnel Simulation of the Atmospheric Surface Layer for the Study of Wind Loads on Low-Rise Buildings," *Journal of Industrial Aerodynamics* 2(3): 21–38, November.

[4] Reinhold, T.A., H.W. Tieleman and F.J. Maher (1978) "Aerodynamic Forces on a Tall Building Model in a Turbulent Boundary Layer," *Experimental Mechanics* 18(6): 201–207, June.

[5] Okada, H. and Y.-C. Ha (1992) "Comparison of Wind Tunnel and Full-Scale Pressure Measurement Tests on the Teaxas Tech Building," *Journal of Wind Engineering and Industrial Aerodynamics* 43: 1601.

[6] Bienkiewicz, B. and H.J. Ham (2003) "Windtunnel Modeling of Roof Pressure and Turbulence Effects on the CTU Building," *Wind and Structures, and International Journal* 6(2): 91.

Index

Note: Page numbers in italics indicate a figure on the corresponding page.

T - #0260 - 111024 - C228 - 254/178/13 - PB - 9780367766191 - Matt Lamination